SPECIAL DISTRICTS IN COOK COUNTY: TOWARD A GEOGRAPHY OF LOCAL GOVERNMENT

Donald Foster Stetzer

University of Wisconsin at Stevens Point

THE UNIVERSITY OF CHICAGO

DEPARTMENT OF GEOGRAPHY

RESEARCH PAPER NO. 169

1975

Library of Congress Cataloging in Publication Data

Stetzer, Donald Foster, 1926–
  Special Districts in Cook County.
  (Research paper—University of Chicago, Department of Geography; no. 169)
  Bibliography: p. 163
  1. Cook Co., Ill.—Politics and government. 2. Special Districts—Cook Co., Ill.
3. Special Districts. I. Title. II. Series: Chicago. University. Dept. of Geography.
Research paper; no. 169.
H31. C514 no. 169 [JS451.I36] 910s [352′.009′097731]
ISBN 0-89065-076-4
75-14240

Research Papers are available from:
The University of Chicago
Department of Geography
5828 S. University Avenue
Chicago, Illinois 60637
Price: $5.00 list; $4.00 series subscription

ACKNOWLEDGMENTS

The gathering of much information for this study has been dependent upon informal interviews with numerous officials of Special Districts, municipalities, and townships in Cook County. It would not be feasible to acknowledge them individually, but without their cooperation and assistance this study would not have been possible. The Illinois Department of Local Government Affairs was very helpful in providing detailed maps by which the actual pattern of Special Districts could be determined.

I wish to thank Dr. Norton Ginsburg for his continuous encouragement throughout the study. His insight, suggestions, and guidance are deeply appreciated. Dr. William Pattison reviewed the final manuscript and made many valuable suggestions for the organization of the study.

Appreciation is expressed to the Department of Geography-Geology at the University of Wisconsin--Stevens Point which provided valuable assistance in preparing the manuscripts. Special thanks are reserved for Mr. Bruce Dustrude who supplied invaluable cartographic assistance throughout the study.

TABLE OF CONTENTS

Previous Research
Organization of the Study
Sources of Data

Chapter

The Special District Defined
    Dependent Districts
    Public Authorities
    School Districts
Literature on Special Districts
History of Special Districts
    Historical Source Materials
Functions of Special Districts
    Urban
    Urban and Non-urban
    Natural Resources
    Miscellaneous
Summary

Systems Analysis
The Easton Framework
Homeostasis
Constraints upon the Political System Leading to the
    Creation of Special Districts
    Financial Limitations
    Limitations on the Powers of Local Government
        Strict Construction of Powers
        Differential Taxing Areas
    Areal Convenience
    Political Compromises
    Desire for Business Management

Characteristics
Legal Aspects
Case Study: South Stickney Sanitary District
Street Light Districts
Summary

The Size of Special Districts
The Distribution of Resources
Horizontal Integration
        Functional Horizontal Integration
        Areal Horizontal Integration
Centralization and Local Control
        The Case for Centralization
        The Case for Local Control
        Summary
An Appraisal of Special Districts

LIST OF TABLES

LIST OF ILLUSTRATIONS

INTRODUCTION

Areal political forms comprise a significant part of the organization of any polity. These forms must be so devised that they come to terms with the needs of the citizens and with the environment of the area within which the polity is imbedded. The spatial organization of these forms is a reflection of the complex interaction of the polity and area--hereafter called the "space-polity."[1] A major objective of this study is to acquire a better understanding of a portion of this space-polity, the role it plays in the larger societal system, how it is related to its environment, and how it came about, so that forecasting its future state will become more rational and informed decisions about deliberate change equally so.

In the United States the individual State governments are empowered to create and allocate powers to local governments within their boundaries. The functions of these local governments may be conveniently divided into two main types: general-purpose and special-purpose. General-purpose governments are authorized to perform and coordinate functions within a legally prescribed area. On the other hand, special-purpose governments are usually empowered to render only a single service within a prescribed area, although this area also lies within the jurisdiction of general-purpose governments.

In most of the United States, the general-purpose governments can be divided into two broad categories: (1) counties[2] and their subdivisions, and (2) municipalities. Counties are territorial subdivisions of the State and are authorized to carry out certain functions in both rural and urban territory. The counties in turn may be divided into smaller units, sometimes collectively referred to as minor civil divisions. These smaller units have various names, but in the

---

[1] The term space-polity is used to denote the spatial dimensions of any political process, institution, or phenomenon. Source: Joseph B. R. Whitney, China: Area, Administration, and Nation Building, Department of Geography Research Paper No. 123 (Chicago: University of Chicago, Department of Geography, 1970), p. 7.

[2] Subdivisions of the State are known as parishes in Louisiana and as boroughs in Alaska.

northeastern and north-central parts of the United States, they are known as towns or townships.

The second class of general-purpose governments consists of municipalities which are authorized to perform a number of functions in densely settled areas. Depending on the State concerned, municipalities may be known as cities, villages, towns, or boroughs.

The special-purpose governments appear in a much wider variety of forms than general-purpose governments. The most important member of this class of governmental forms is the Special District. The characteristics and problems of definition of this form are complex and differ widely from State to State. A more precise and comprehensive definition of the Special District will be given later.

This study will concentrate on the development of Special Districts in Cook County, Illinois. This densely populated county contains the city of Chicago and a multitude of smaller municipalities. The long history of Special District government in the county, the complexity of its governmental forms, and its differing stages of urbanization afford a great potential for studying Special Districts in many situations.

The immediate goals of this study are (1) to develop a conceptual framework within which Special Districts can be placed; (2) to identify the factors and processes in Special District development; (3) to describe the functions and areal organization of Special District government; and (4) to assess the relationship of Special Districts to other aspects of government.

Previous Research

Most studies in political geography have been concentrated on the nation-state as the primary object of analysis; although, from time to time, studies have appeared on local administrative areas as related to subdivisions of the nation-state. Still, in relation to the importance of the subject, there has been comparatively little work on it, although some information on local governmental areas is available as byproducts from boundary studies.[1]

---

[1] George F. Brightman, "The Boundaries of Utah," Economic Geography, XVI (January, 1940), 87-95; E. W. Gilbert, "The Boundaries of Local Government Areas," Geographical Journal, CXI (April-June, 1948), 172-206; Lawrence Martin, "The Michigan-Wisconsin Boundary Case in the Supreme Court of the United States, 1923-1926," Annals of the Association of American Geographers, XX (September, 1930), 105-83 and "The Second Wisconsin-Michigan Boundary Case in the Supreme Court of the United States, 1932-1936," Annals of the Asso-

Some information on local governmental areas is also available in studies of particular areas.[1] It is difficult to make any generalizations about these studies because they lack comparability and deal with very different physical and cultural environments. Most of the studies on Special Districts are in the field of local government, and geographic studies are relatively few.[2]

Since the study of Special Districts is so specialized and seemingly unrelated to conventional geographic research, a discussion of the literature in cognate fields will be deferred to Chapter I, where a fuller discussion of the Special District as a governmental form appears.

---

ciation of American Geographers, XXVIII (March, 1938), 77-126; Benjamin E. Thomas, "The California-Nevada Boundary," Annals of the Association of American Geographers, XLII (March, 1952), 51-68; and Edward L. Ullman, "The Eastern Rhode Island-Massachusetts Boundary Zone," Geographical Review, XXIX (April, 1939), 291-302. For further work on boundaries, see: Julian V. Minghi, "Boundary Studies in Political Geography," Annals of the Association of American Geographers, LIII (September, 1963), 407-28; also J. R. V. Prescott, The Geography of Frontiers and Boundaries (Chicago: Aldine Publishing Co., 1965).

[1]Sherwin H. Cooper, "The Census County Division: A Major Revision of the Minor Civil Division," Professional Geographer, XV (July, 1963), 4-8; Harm J. de Blij and Donald L. Capone, "Wildlife Conservation Areas in East Africa: An Application of Field Theory in Political Geography," Southeastern Geographer, IX (November, 1969), 94-107; John D. Eyre, "City-County Territorial Competition: The Portsmouth, Virginia Case," Southeastern Geographer, IX (November, 1969), 26-38; John H. Garland, "The Western Reserve of Connecticut: Geography of a Political Relic," Economic Geography, XIX (July, 1943), 301-19; J. Malcolm Holmes, The Geographical Basis of Government, Specially Applied to New South Wales (Sydney: Angus and Robertson, Ltd., 1944); Roger E. Kasperson, "Toward a Geography of Urban Politics: Chicago, A Case Study," Economic Geography, XLI (April, 1965), 95-107; Roger E. Kasperson, "Environmental Stress and the Municipal Political System," in The Structure of Political Geography, ed. by Roger E. Kasperson and Julian V. Minghi (Chicago: Aldine Publishing Co., 1969), pp. 481-96; Alexander Melamid, "Municipal Quasi-Exclaves: Examples from Yonkers, N. Y.," Professional Geographer, XVIII (March, 1966), 94-96; Raymond E. Murphy, "Town Structure and Urban Concepts in New England," Professional Geographer, XVI (March, 1964), 1-6; Howard J. Nelson, "The Vernon Area, California--A Study of the Political Factor in Urban Geography," Annals of the Association of American Geographers, XLII (June, 1952), 177-91; Malcolm J. Proudfoot, "Chicago's Fragmented Political Structure," Geographical Review, XLVII (January, 1957), 106-17; Griffith Taylor, "Towns and Townships in Southern Ontario," Economic Geography, XXI (January, 1945), 88-96.

[2]David C. Winslow, "Geographical Implications of Soil Conservation Districts in the United States," Professional Geographer, I (May, 1949), 11-14. References to Special Districts in geographic books are: W. A. Douglas Jackson and Edward F. Bergman, A Geography of Politics (Dubuque, Iowa: William C. Brown Co., 1973), pp. 22-36, passim; Raymond E. Murphy, The American City: An Urban Geography (New York: McGraw-Hill Book Co., 1966), pp. 427-28; and Norman J. G. Pounds, Political Geography (2nd ed.; New York: McGraw-Hill Book Co., 1972), pp. 224-30.

In addition to the concentration on the nation-state, much political geographic research has had its research emphasis on physical factors such as topography and climate; but slowly this early emphasis has given way to cultural factors such as values, traditions, and political power. Robert Platt, for example, recognized that areas of organization were generally independent of terrain factors and that these areas were not always visible in the physical landscape. He recognized that long periods of association in a particular area led to organizational patterns that resisted change.[1] Another example of the recognition of cultural factors is found in Whitney's study of administrative areas in China.[2] In this work, he carefully examined the influence of the factors of political control and the process of national development on the size and number of administrative subdivisions.

Under the direction of Platt, a comparative study was made of a community of Chicago (Mount Greenwood) and an adjoining independent suburb (Evergreen Park).[3] In this study, Olson found that there was little difference in land use or socioeconomic characteristics between the two areas. The chief differences were in functional organization, and these differences had developed through the interaction of political groups over a long period of time.

Another study in an urban area was Nelson's study of Vernon, California --then a part of the Los Angeles urban fringe.[4] Unlike Olson, he found that significant land-use differences between adjoining areas had arisen through functional differentiation. In contrast to the surrounding residential areas of Los Angeles, Vernon had a large proportion of its land devoted to industry, resulting from a deliberate policy of attracting industry on the part of the decision-makers in Vernon.

Both of these studies flowed out of an old tradition in geography--the description of a landscape complex and then identifying the factors that influenced its formation. The significant element in both studies was the change in areal organization and accompanying change in functions when a political boun-

---

[1] Robert S. Platt, A Geographical Study of the Dutch-German Border (Münster-Westfalen, W. Germany: Geographische Kommission, 1958), pp. 69-71.

[2] Whitney, China: Area, Administration, and Nation Building, pp. 80-88, 166-72.

[3] Howard E. Olson, "Evergreen Park and Mount Greenwood Astride Chicago's Boundary" (unpublished M.A. dissertation, Department of Geography, University of Chicago, 1954).

[4] Nelson, "Vernon Area, California," pp. 177-91.

dary was crossed. However, it was shown that a change in political organization may or may not have been accompanied by a change in the landscape.

In a move away from concentration on the landscape, but still within the Platt tradition, Brown applied the concept of a hierarchical ordering in economic organization to political areas.[1] He studied all the levels of government that existed in the Saint Cloud, Minnesota area and placed them in an administrative hierarchy. In his study, single-purpose Special Districts were in the lowest level of the hierarchy of governments.

Organization of the Study

From the foregoing discussion of geographic research, it is apparent that much groundwork must be laid before an interpretation of Special Districts can be made. As a beginning, Chapter I of this study contains a general discussion of Special Districts in the United States. The relevant literature is reviewed, and considerable attention is paid to the historical development of these forms and to the complex terminology used in describing them.

Chapter II is concerned with the development of a theoretical framework in which Special Districts may be placed. Using some of the concepts from systems analysis, Special Districts are shown to be homeostatic mechanisms resulting from the interaction of the political system with its environment.

Departing from the general topics covered in the first two chapters, Chapter III deals with the specific geographical and historical setting of Special Districts in Cook County. Special Districts are interrelated with counties, townships, and municipalities, and an understanding of Special Districts cannot be obtained without a general knowledge of the history, functions, and location of these other governmental forms.

Chapters IV, V, and VI deal with the three main types of districts in Cook County. These districts are grouped by their differing relationships with the general-purpose governments over which they are superimposed. Chapter IV contains a survey of the districts that cover extensive areas and are known as "area-wide districts." These districts tend to follow county and township boundaries, and areas within these districts are subject to both municipal and nonmunicipal governments. Chapter V is concerned with districts that serve primarily municipalities. These districts provide services that appear "municipal"

---

[1]Robert Harold Brown, Political Areal Functional Organization: With Special Reference to St. Cloud, Minnesota, Department of Geography Research Paper No. 51 (Chicago: University of Chicago, 1957).

in character, but the services are provided by legally separate governments. Chapter VI is devoted to districts that serve areas which are urban in some respects, although they lack municipal government. Districts of this type are called "fringe districts" in this study because of their location on the edges of municipalities.

Chapter VII concludes the study, dealing with the impact of Special Districts on five selected aspects of the space-polity of Cook County: (1) the size of Special Districts; (2) the distribution of resources; (3) horizontal integration; (4) centralization and local control; and (5) an appraisal of Special Districts.

Sources of Data

Data have been collected from a wide variety of sources which vary greatly in quality and availability. Each type of district, and indeed each district itself, has a unique historical and legal background, greatly affecting the type of information that can be obtained. Special reports, legal documents, census data, maps, newspapers, and pamphlets were examined in libraries of municipalities and academic institutions.

For every type of district, the data are lacking in some particulars, necessitating much reconstruction of fragmentary evidence into a coherent whole. The maps and records in the offices of the Cook County Government are for all practical purposes inaccessible, requiring extensive contacts with local governments and with the State of Illinois.

The documentary sources were supplemented by dozens of interviews with public officials and governmental employees who manifested varying degrees of cooperation. Thus, the data employed consisted of literally hundreds of bits of information, normative statements, and inferences, the accumulation, evaluation, and interpretation of which has occupied the better part of five years.

CHAPTER I

INTRODUCTION TO THE SPECIAL DISTRICT

An important feature of local government in the United States is the mosaic of Special Districts that characterizes the political landscape. Compared to the traditional governmental forms, these special governmental forms constitute a long neglected part of political and urban geography. As Bollens remarks, "Special Districts, particularly those in the nonschool categories, constitute the new dark continent of American politics, a phrase applied earlier in the century to counties."[1]

The Special District Defined

Special Districts are a class of organized governmental entities that have a structure, an official name, perpetual succession, the power to perform certain functions, the right to sue and be sued, the right to make contracts, and the right to dispose of property. They have officials chosen in various ways and possess considerable fiscal and administrative independence.[2]

The Special District is a significant member of that group of governments which exists outside the traditional "line" structure of nations, states, counties, townships, and municipalities. In addition to Special Districts, these non-traditional governments include such forms as authorities, boards, agencies, commissions, and corporations.

The term Special District is used in both a general sense and in a restricted sense. Through time, a gradual classification of Special Districts has taken place, and a more precise terminology has evolved. The terms "dependent districts," "public authorities," and "school districts" are used increas-

---

[1] John C. Bollens, Special District Governments in the United States (Berkeley: University of California Press, 1961), p. 1.

[2] Ibid.

7

ingly often by contemporary writers to refer to Special Districts with specific characteristics. By splitting off these specific types, the term Special District, as currently employed, has acquired a more restricted meaning and simply means that it is not one of the other types.

Unfortunately, the name of a governmental unit is not a reliable guide to its true character, and a more precise definition may be required. To make matters more complicated, all writers do not use the same definitions or terminology. Before proceeding further, then, it is necessary to examine carefully the terminology that is in common use.

The United States Bureau of the Census (hereafter referred to as the Census Bureau) has done extensive work in classifying governmental forms. It uses the general term "Special District" to include many types of governments that may not necessarily have the word "district" in their names. In its standards for counting governmental units, the Census Bureau stresses independence in operations and policy-making. It states:

> A government is an organized entity which, in addition to having governmental character, has sufficient discretion in the management of its own affairs to distinguish it as separate from the administrative structure of any other governmental unit.

> To be counted as a government, any entity must possess all three of the attributes reflected in the foregoing definitions: Existence as an organized entity, governmental character, and substantial autonomy. [1]

Dependent Districts. --For governments that are not independent of other governments and not listed by the Census Bureau, Bollens uses the term "dependent districts." As Bollens describes these forms:

> Some entities that are not independent special districts resemble them in some particulars and are at times mistaken for them. . . . they deserve separate consideration principally for purposes of clarification and also because they are increasing in number and importance. The fundamental distinction between such operations and special district governments is the former's lack of sufficient fiscal independence or inadequate administrative autonomy or both. [2]

The term dependent district is not widespread, and most writers make no distinction between dependent and independent Special Districts. The reasons for such a lack of distinction is made clear by Bollens:

---

[1] U.S. Bureau of the Census, Census of Governments: 1967, Vol. I: Governmental Organization, p. 13.

[2] Bollens, Special District Governments, p. 228.

It is easy to confuse some dependent agencies and operations with special districts that are independent governments. Many of them have the same functions and names as special districts. There are perplexing interstate differences. Although housing authorities, soil conservation districts, and school districts are generally independent governments, each category is subordinate in some states.[1]

Public Authorities. --The terms "public authority" and Special District have been used interchangeably by many writers. For example, the National Municipal League in 1930 used the term "special metropolitan authorities" for all types of governments, other than traditional forms, that operated in metropolitan areas.[2] The League used the terms "districts" and "authorities" without differentiating them. In its tabulation of governments, the Census Bureau considers independent public authorities as Special Districts.

In recent years, the term "public authority" has been restricted to a particular type of Special District--that is, one that issues bonds to be repaid by user charges. As Smith puts it:

> Special districts, or public authorities, being as they are, ad hoc efforts to meet peculiar contingencies, almost defy description. The one feature, however, which commonly reoccurs in attempts at definition is that of bond-revenue method of finance.[3]

Smith does not use the criterion of administrative independence in separating out public authorities from the more general class of Special Districts. On the other hand, Bollens tries to include both independence and methods of financing in his discussion:

> Dependent districts and authorities are not always clearly distinguishable, and the titles have sometimes been employed interchangeably by state legislatures. Authorities, however, more frequently engage in revenue-producing enterprises financed solely by revenue bonds and service charges and rates, with or without support from governmental grants.[4]

Although there are many exceptions, Bollens makes the observation that authorities are generally adjuncts of other governments and that districts generally tend to be independent governments.[5]

---

[1] Ibid., p. 229.

[2] National Municipal League, The Government of Metropolitan Areas in the United States (New York: National Municipal League, 1930), p. 256.

[3] Robert G. Smith, Public Authorities, Special Districts and Local Government: A Digest of Excerpts (Washington: National Association of Counties Research Foundation, 1964), p. 20.

[4] Bollens, Special District Governments, p. 229.

[5] Ibid., p. 231.

School Districts. --The term school district long has been recognized as a type of Special District. Recent studies of Special Districts, however, have omitted school districts from their discussions, and the term Special District is evolving to mean non-school districts. The Census Bureau includes school districts in its publications on Special Districts, although school districts are listed separately. As an example of this modern usage, the 1970 Illinois Constitution refers to Special Districts and school districts as separate classes of governments.[1]

Literature on Special Districts

As was noted in the Introduction, very little information on Special Districts appears in the literature of geography.[2] Most of the literature is in the fields of public administration, local government, law, and finance. The bulk of this information concerns statutory provisions, legislative history, court decisions, process of formation, selection of officials, techniques of raising revenues, and debt structure.

Owing to the great diversity of Special Districts and large gaps in information, only a few studies generalize about them at the national level.[3] About one-third of the States are covered by studies generalizing at the State level.[4]

---

[1] Illinois, Constitution (1970), art. vii, sec. 1.

[2] The best short account of Special Districts in the geographic literature appears in Jackson and Bergman, Geography of Politics, pp. 22-36, passim. Another good account is found in Pounds, Political Geography, pp. 224-30.

[3] The classic work in this field is the previously cited Bollens, Special District Governments. Also at the national scale are Advisory Commission on Intergovernmental Relations, The Problem of Special Districts in American Government (Washington: U.S. Government Printing Office, 1964) and Robert G. Smith, Public Authorities, Special Districts and Local Government (Washington: National Association of Counties Research Foundation, 1964). The best bibliography on Special Districts is U.S. Department of Agriculture, Economic Research Service, A Selected Bibliography on Special Districts and Authorities in the United States, Annotated, Miscellaneous Publication No. 1087 (Washington: U.S. Government Printing Office, 1968).

[4] The following sources are considered to be the best works on a single State. They are arranged alphabetically by States. Estal E. Sparlin, Special Improvement District Finance in Arkansas, Agricultural Experiment Station Bulletin No. 424 (Fayetteville, Ark.: University of Arkansas, College of Agriculture, 1942); Stanley Scott and John C. Bollens, Special Districts in California Local Government (Berkeley: University of California, Bureau of Public Administration, 1949); Clyde F. Snider and Roy Anderson, Local Taxing Units: The Illinois Experience (Urbana: University of Illinois, Institute of Government and

State legislation and court decisions tend to enforce a uniformity on Special Districts within a single State, particularly in regard to selection of officials, methods of financing, authorization of powers, and procedures for annexation and dissolution.

Studies of Special Districts below the statewide level of generalization are extremely varied.[1] They tend to be narrowly focused and deal with limited

---

Public Affairs, 1968); William H. Cape, Leon B. Graves, and Burton M. Michaels, Government by Special Districts, Kansas Governmental Research Series No. 37 (Lawrence, Kan.: University of Kansas, Governmental Research Center, 1969); Harvey Lee Waters and William H. Raines, Special Districts in Kentucky, Research Report No. 48 (Frankfort, Ky.: Legislative Research Commission, 1968); Emmett Asseff, Special Districts in Louisiana (Baton Rouge: Louisiana State University, Bureau of Government Research, 1951); Richard Folmar, Special District Governments in New Mexico (Santa Fe: New Mexico Legislative Council, 1962); Pennsylvania Government Administration Service, Municipal Authorities in Pennsylvania (Philadelphia: Pennsylvania Government Administration Service, 1941); Frederick L. Bird, Local Special Districts and Authorities in Rhode Island, Research Series No. 4 (Kingston, R.I.: University of Rhode Island, Bureau of Government Research, 1962); Woodward G. Thrombley, Special Districts and Authorities in Texas (Austin: University of Texas, Institute of Public Affairs, 1959); J. D. Williams, A Report on Utah's Special Purpose Districts (Salt Lake City: University of Utah, Institute of Government, 1957); S. J. Makielski, Jr. and David G. Temple, Special District Government in Virginia (Charlottesville, Va.: University of Virginia, Institute of Government, 1967); University of Washington, Bureau of Governmental Research and Services, Special Districts in the State of Washington, Report No. 150 (Seattle: University of Washington, Bureau of Governmental Research and Services, 1963); Joseph Geraud, Special Taxing Districts in Wyoming, A Report to the Wyoming Legislative Research Committee, Research Report No. 5 (Cheyenne, 1960).

[1]Good studies that deal with a single class of districts are: Henry Bain, The Development District: A Governmental Institution for the Better Organization of the Urban Development Process in the Bi-County Region, Report for the Maryland-National Capital Park and Planning Commission (Riverdale, Md.: Washington Center for Metropolitan Studies, 1968); Donald J. Brosz, Establishing Water Conservancy Districts in Wyoming, Bulletin No. 530 (Laramie, Wyo.: University of Wyoming, Agricultural Extension Service, 1970); Missouri State Soil and Water Districts Commission, Organization and Operation of Soil and Water Conservation Districts in Missouri, Circular No. 1 (Columbia, Mo., 1964); Public Affairs Research Council of Louisiana, Louisiana Levee Districts (Baton Rouge, La., 1958); Arthur B. Winter, The Tennessee Utility District: A Problem of Urbanization (Knoxville, Tenn.: University of Tennessee Press, 1958); Wisconsin Taxpayer, "Special Sanitary Districts," The Wisconsin Taxpayer, Vol. XXXVIII (October, 1970).
    Examples of detailed studies that deal with micro-areas include: Lloyd K. Fischer and John F. Timmons, Progress and Problems in the Iowa Soil Conservation Districts Program: A Pilot Study of the Jasper Soil Conservation District, Research Bulletin No. 466 (Ames, Iowa: Iowa State College, Agricultural and Home Economics Experiment Station, 1959); Oregon, State Engineer, Final Report on the Advisability of Creating the Josephine County People's Utility District (Salem, Ore., 1962); Park College, Governmental Research Bureau, The

aspects of district development. Moreover, most are not referenced in standard bibliographic sources and can be found only in specialized collections.

History of Special Districts

Having defined the Special District, as well as cited the relevant literature, we now are ready to provide a brief historical account of the Special District as a governmental form. For convenience, this account will be confined to England and the United States, although it is probable that Special Districts (or analogous forms) have developed in other countries.[1]

Beginning in the 1600's in England, the traditional governmental forms of counties, parishes, and boroughs became highly rigid and increasingly inadequate for solving problems of public health and transportation arising from greater urbanization and the wider use of the stagecoach. A system of ad hoc authorities worked with the then new urban problems of paving, lighting, policing, and cleaning of streets. Turnpike trusts were created to meet the increased demand for new roads. After these forms had reached their high point in the nineteenth century, a change in public attitudes began to take place. Municipali-

Special Districts of Platte County, Missouri, Parts I and II (Parkville, Mo., 1958).

Two good studies dealing with Special Districts lying adjacent to cities are: Stanley Scott and John Corzine, Special Districts in the San Francisco Bay Area: Some Problems and Issues (Berkeley, Cal.: University of California, Institute of Governmental Studies, 1963) and University of Oregon, Bureau of Municipal Research and Service, Problems of the Urban Fringe, Vols. I and II (Eugene, Ore., 1957).

Two of the best studies that deal with the general topic of Special Districts in urban areas are: Max A. Pock, Independent Special Districts: A Solution to the Metropolitan Area Problem (Ann Arbor: University of Michigan Law School, 1962) and Robert G. Smith, Public Authorities in Urban Areas (Washington: National Association of Counties Research Foundation, 1969).

Studies that deal with Special District organization are: Arthur W. Bromage, Political Representation in Metropolitan Agencies, Michigan Governmental Studies No. 42 (Ann Arbor: University of Michigan, Institute for Public Administration, 1962); Institute for Local Self Government, Special Districts or Special Dynasties?: Democracy Diminished (Berkeley: Institute for Local Self Government, 1970); Nathaniel Stone Preston, Public Authorities: Devices to Finance, Construct, and Operate Public Projects: How Do They Work? Grass Roots Guides on Democracy and Practical Politics, Booklet No. 40 (Washington: Center for Information on America, 1969); New Jersey State Chamber of Commerce, Department of Governmental and Economic Research, Government by "Authorities" for New Jersey? (Newark, 1952).

[1]In the next chapter, an account is given concerning community land-grant districts which developed in New Mexico while the area was still under Spanish and Mexican rule.

13

ties were expected, as a matter of course, to provide urban services. Similarly, county and borough governments were expected to build and maintain roads for public use. As Winter says, "Although there still remain many independent local authorities in Britain, the majority of civil governmental agencies have come within the purview of regular municipal and county borough jurisdiction."[1]

In the early settlement of the United States, many legal and governmental forms of English origin were adopted by the American colonies. In describing early Special Districts in Virginia, Makielski and Temple observe:

> Special districts were used extensively in Virginia prior to 1700. The special purpose district probably began with the administration of the Elizabethan Poor Law in 1621 and the later creation of the Overseers of the Poor. By the time of the Revolution, special commissions were created to build and repair roads, erect bridges, and construct and maintain tobacco warehouses. Nearly every stream in Virginia was the province of an ad hoc body which was charged with making the watercourse navigable.[2]

Throughout the early history of the United States, the toll road and canal corporations of the early 1800's were an early use of Special Districts. Other Special Districts were created to provide benefits to limited groups of property owners for maintaining local roads or providing protection against fires and floods. The district generally provided an extremely limited service, benefited a small group of people, and rarely interfered with programs of general-purpose local governments.[3]

As early as 1789 school districts were established in Massachusetts, and during the 1800's and early 1900's thousands of school districts were established in the United States. There was considerable sentiment for establishing separate governments for educational purposes; for the people believed that separation from old-line units of government would provide insulation and safety from "dubious and sinister political pressures."[4]

In metropolitan areas, specialized-function districts have been used since colonial times. For example, beginning in 1790 in Philadelphia, special authorities were created for the administration of prisons, poor relief, port development, public health control, police, and education. In 1865 New York

---

[1]Winter, Tennessee Utility District, p. 7.

[2]Makielski and Temple, Special District Government in Virginia, p. 10.

[3]Advisory Commission on Intergovernmental Relations, Special Districts in American Government, p. 1.

[4]Winter, Tennessee Utility District, p. 7.

(Manhattan Island) and Brooklyn were constituted a single metropolitan fire district. In 1869 park districts were created for Chicago and adjoining townships.[1]

Through the latter half of the nineteenth century and early decades of the twentieth century, the number of Special Districts grew slowly but steadily. It was not until the depression of the 1930's that their rate of growth was to suddenly increase. There was a drastic reduction in State financial resources, and revenue bonds and authority financing became a popular mechanism for supporting public works. In 1934 President Roosevelt recommended that each State pass enabling legislation to create public bodies with power to issue bonds to finance revenue-producing improvements. These public bodies permitted local participation in Federal programs.[2] In response to this participation, public housing authorities and soil conservation districts multiplied through the country after 1937.

The scarcity of equipment and materials for capital construction purposes during World War II proved to have a dampening effect on the growth of Special Districts. The removal of these limitations, however, after the war provided the impetus for the phenomenal growth in the number of Special Districts. The country's rapidly expanding population, benefiting from post-war prosperity and wider use of the automobile, brought about extensive urban development in the areas adjacent to cities. The rapid increase in population generated sudden demands for greatly increased services. The existing governmental structure was frequently unresponsive to these new demands, and Special Districts were created to provide additional services. As Makielski and Temple observe:

> The developed areas required services: water, sewer lines, parks and playgrounds, street lights, garbage removal, and sidewalks. Local special districts proliferated to meet the need for services. Water authorities, sewer authorities, airport authorities, parking authorities, and park authorities appeared on the Virginia scene.[3]

Table 1 shows the development of units of local government in the United States during the last thirty years. The traditional forms--counties and townships--have remained fairly stable, and the number of municipalities has increased slowly. School districts have dropped dramatically in number as much

---

[1] National Municipal League, Government of Metropolitan Areas, pp. 257-59.

[2] Robert Gerwig, "Public Authorities: Legislative Panacea?" Journal of Public Law (January, 1957), pp. 387-88.

[3] Makielski and Temple, Special District Government in Virginia, p. 27.

TABLE 1

UNITS OF LOCAL GOVERNMENT IN THE UNITED STATES, 1942-1972

	1942	1952	1957	1962	1967	1972
Counties	3,050	3,052	3,050	3,043	3,049	3,044
Townships	18,919	17,202	17,198	17,142	17,105	16,991
Municipalities	16,220	16,807	17,215	18,000	18,048	18,517
School districts	108,579	67,355	50,454	34,678	21,782	15,781
Non-school Special Districts	8,299	12,340	14,424	18,322	21,264	23,885
Total	155,067	116,756	102,341	91,186	81,248	78,218

Sources: U.S. Bureau of the Census, Census of Governments: 1967, I, 23, and Census of Governments: 1972, I, 1.

consolidation has taken place. In contrast, non-school Special Districts have increased more rapidly than any other form.

Historical Source Materials. --Much of the history of Special Districts remains to be developed. The Census Bureau has maintained a count of Special Districts only since 1942. In recent years, the story of the historical development of Special Districts is slowly emerging as more and more research is done in individual States. Nevertheless, the historical record of Special Districts in the majority of States still remains to be uncovered. In many areas, an abysmal lack of records and the absence of any "archival sense" makes necessary a laborious and frustrating search of incomplete records, complex laws, and fragmentary accounts to piece together the actual events that occurred.

Between 1930 and 1940, William Anderson began the monumental task of systematically gathering information on the local governmental units of the United States. His comments about gaps in information reveal some of the reasons for the slowness in developing studies on Special Districts. As he comments:

> Hardest of all to deal with are the special districts other than school districts. This difficulty can be explained partly as follows: (1) No state agency is responsible for compiling and reporting the information concerning them, nor need they in most cases record with any state or county authority the fact of their organization. (2) As a rule they do not levy direct property taxes; in consequence, the state and county tax authorities are practically unaware of their existence. (3) New classes of such units are easily and frequently authorized by state legislation without much public notice. It would be necessary to read all the new legislation of every state each year in order to get on the trail of all these new units. (4) Under such laws local units are rather easily suspended or dissolved, and the facts are not noted anywhere. (5) Many such local units are distinctly borderline cases. Whether they are to be classed as separate units of government or not is often a disputable question. [1]

Functions of Special Districts

The functions of non-school Special Districts encompass an amazing variety of activities. Most districts perform only a single function, and the function is indicated by the name of the district. In a few States, multi-purpose Special Districts have been created. [2]

---

[1] William Anderson, The Units of Government in the United States (rev. ed.; Chicago: Public Administration Service, 1949), p. 5.

[2] These multi-purpose Special Districts sometimes are called "junior city districts." They are limited to fewer powers than a general-purpose gov-

The Advisory Commission on Intergovernmental Relations has classi-
fied the functions of Special Districts into four categories: (1) urban; (2) mixed
urban and non-urban; (3) natural resources; and (4) miscellaneous.[1] This classification in part is based upon Special Districts recognized as independent governments by the Census Bureau (Table 2). Great care must be exercised in
using the Census Bureau figures because the same type of district may be independent in one State (and counted), but in another State the district may not be
independent (and not counted).

TABLE 2

NON-SCHOOL SPECIAL DISTRICTS
IN THE UNITED STATES, 1972

Function	Number
Single-function:	
Natural resources:	
Soil conservation	2,561
Drainage	2,192
Irrigation, water conservation	971
Flood control	684
Other and composite resource purposes	231
Fire protection	3,872
Urban water supply	2,333
Housing	2,271
Cemeteries	1,494
Sewerage	1,411
School buildings	1,085
Parks and recreation	750
Highways	698
Hospitals	657
Libraries	498
Other single-function districts	1,273
Multiple-function districts	904
Total	23,885

Source: U.S. Bureau of the Census, Census of Governments: 1972, I, 5.

ernment, such as a city; but they are functionally much broader than a single-
purpose Special District. For a fuller discussion, see Bollens, Special District
Governments, pp. 106-9.

[1]Advisory Commission on Intergovernmental Relations, The Problem of
Special Districts in American Government, p. 12.

Urban. --The districts of this type are predominantly in urban areas, although rural territory sometimes is included. Districts in this category include fire protection, housing, urban renewal, parks, recreation, sewage disposal, water supply, electric power, gas supply, mass transit, ports, airports, and street lighting.

Urban and Non-urban. --These districts are not confined to either urban or rural areas. They provide services for and the maintenance of public health, hospitals, libraries, roads, and cemeteries.

Natural Resources. --These districts, confined largely to rural areas, include soil conservation, drainage, irrigation, water conservation, flood control, and levees.

Miscellaneous. --For the most part, these districts are not enumerated by the Census Bureau since they are considered as subordinate to other governments.[1] In rural areas, districts of this type exist for the protection from and the control of noxious weeds, predatory animals, harmful birds, and pestiferous insects. Other districts in this category include such varied activities as forest preserves, parking lots, county fairs, fish and game management, air pollution, bridges, mining, and television.

Summary

In this chapter we have reviewed the development of Special Districts as governmental forms and clarified some of the complex terminology used in describing them. The existing literature has been compiled, and the amazing variety of functions performed by Special Districts has been described.

We now might logically ask, "Why has all of this happened?" The following chapter endeavors to answer this question by developing a conceptual framework within which the role of the Special District in a political system might be explained. Once this framework is developed, we then shall attempt to identify the individual elements that lead to the formation of Special Districts.

---

[1]U. S. Bureau of the Census, Census of Governments: 1967, I, 297-457, passim. At the end of a discussion of the governmental organization for each State, the Census Bureau lists some of the types of districts that it does not consider as being independent Special Districts.

CHAPTER II

THE ROLE OF SPECIAL DISTRICTS

IN THE POLITICAL SYSTEM

Geographers have done relatively little work on the creation of local administrative areas. Empirical studies of the differentiation of political areal forms at the local level are few. Theoretical rationales or models for expressing the patterns of decisions made in politically organized space are just beginning to be developed.[1]

Unfortunately, none of these models appears to be sufficiently encompassing to satisfactorily explain the role of Special Districts in the political system. Accordingly, it will be necessary to first develop concepts and frameworks which will permit the categorization of the thousands of individual actions that constitute political behavior. As a means of developing this framework, the concept of systems analysis will be introduced.

Systems Analysis

Rapoport defines a system as (1) something consisting of a set (finite or infinite) of entities (2) among which a set of relations is specified so that (3) deductions are possible from some relations to others and from the relations among the entities to the behavior or history of the system.[2] Within this broad

---

[1] Four models used by geographers for the political organization of area are: (1) Stephen B. Jones, "A Unified Field Theory of Political Geography," Annals of the Association of American Geographers, XLIV (June, 1954), 111-23; (2) Kasperson, "Environmental Stress and the Municipal Political System," in Structure of Political Geography, ed. by Kasperson and Minghi, p. 485; (3) Edward W. Soja, The Political Organization of Space, Commission on College Geography Research Paper No. 8 (Washington: Association of American Geographers, 1971), p. 7; (4) Bryan H. Massam, The Spatial Structure of Administrative Systems, Commission on College Geography Resource Paper No. 12 (Washington: Association of American Geographers, 1972), pp. 3-26.

[2] Anatol Rapoport, "General Systems Theory," International Encyclopedia of the Social Sciences, XV (1968), 453.

definition, we can select related elements and call them a system--e.g. social system, psychological system, or political system. A system then can be broken down into smaller units called sub-systems which change so slowly that they may be assumed to be temporarily constant. Such an assumption facilitates discussion of the interaction among the sub-systems as well as the change within a single sub-system.

In this study we shall use Easton's systems analytic framework for analyzing the political system. Once certain basic concepts are developed, we shall proceed to develop the systems-related concepts of the steady state and homeostasis and show how these concepts may be applied to the study of Special Districts. Finally, armed with a conceptual framework, we shall examine the constraints upon the political system that lead to the formation of Special Districts.

The Easton Framework

The analysis of the political system by David Easton throws light on the nature of the political process in its environmental setting.[1] In Easton's conception, a political system is defined as a selected set of interactions abstracted from the totality of social behavior, through which values are authoritatively allocated for a society. Those aspects which exist outside the political system constitute the social environment of the political system. Easton then combines this newly defined social environment with other environments such as physical, cultural, and economic into the total environment. The spatial aspects of the interactions of the political system with the total environment provide a foundation for much of contemporary political geography.

In examining Easton's framework further, we may characterize the political system as a self-regulating system dependent on the operation of four functionally different sub-systems: (1) a sub-system of inputs; (2) a sub-system that processes and converts inputs into outputs; (3) a sub-system of outputs; and (4) a feedback mechanism by which the outputs modify the new inputs coming into the system or which may be relayed to the conversion section for modifying subsequent outputs. This system is an open system in that it continually interacts with the environment. See Figure 1 for a diagram of the framework.

The sub-system of inputs consists of two analytically separate elements:

---

[1] David Easton, A Framework for Political Analysis (Englewood Cliffs, N.J.: Prentice-Hall, Inc., 1965) and A Systems Analysis of Political Life (New York: John Wiley & Sons, 1965).

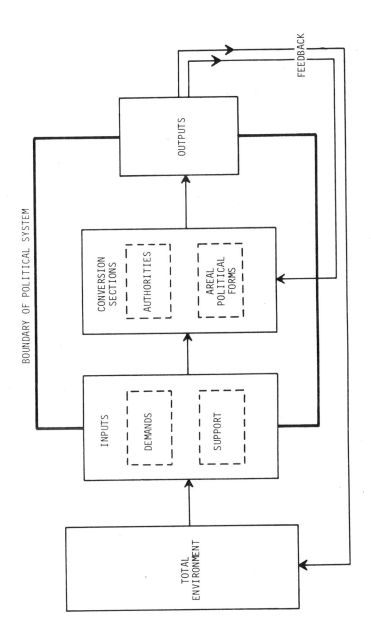

Fig. 1.--Framework for Systems Analysis of Political Activity

demands and support. The environment is continually changing and causes people to make new demands on the political system. Thus we may define demands as those perceived needs of people that have been sufficiently articulated to have found their way to the conversion section. The input of support is the second principal link between the environment and the political system. This continued support is essential for the continuous operation of the conversion section. Support is indicated by such actions as voting, payment of taxes, serving on committees, and obedience to laws and ordinances.

The conversion section contains the governmental and party control structures within which fall the activities of innumerable authorities that produce the outputs of government. The authorities include a wide variety of people, such as officials, trustees, party leaders, legislators, aldermen, and judges. Many of the authorities may comprise the leadership of local areal political forms such as counties, townships, municipalities, and Special Districts; but some authorities--e.g. legislators and judges--have a broader territorial base and are associated with State and National governments.

All demands and estimates of support are fed into the conversion section where they are aggregated, weighed, and interrelated to produce a set of outputs. It is this section that determines the amount of energy and resources that are used in establishing priorities, deciding the quality of services, planning for the future, and distributing rewards.

The outputs of the political system are shaped by the interactions within the conversion section, although they are actually produced by the governmental structure. Outputs include construction, services, regulations, and operations for the well-being and safety of the citizens. These outputs are the manifestations of governmental output that reaches the public, and they are essentially the stakes for which participants in the political system are competing.

An important link in the whole political system is the feedback mechanism, since it is through feedback that the system is continuously modified, enabling it to survive. Stresses and strains brought about by impacts from the environment become noticed by the authorities. They, in turn, adapt the system to fulfill demands so that continued support for the system is forthcoming.

Homeostasis

The Easton framework shows how the political system operates, but it lacks a sense of ongoing process that would tell us why the system operates as it does. Hence, we must build additional characteristics into the framework in order to analyze dynamic political processes as they occur through time. Accordingly, the concept of homeostasis will be developed in order to provide both historical and predictive dimensions.

Homeostasis is a concept widely used in physiology and psychology to identify the tendency of living organisms to maintain and restore steady states of the organisms. The concept also may be extended to apply to a complex social phenomenon such as a political system. Like a thermostat in a heating system, the political system also uses mechanisms to restore the steady state. In order to apply this concept to a complex system, however, we first must identify the steady state, the factors disrupting the steady state, and the mechanisms to restore the steady state.[1]

By their very nature, political systems are open systems that tend to persist or achieve a steady state over time. Steady state does not mean that the structure of the system does not change. On the contrary, it is a device to portray the capability of a system to evolve in structure in order to perpetuate itself or maintain certain crucial system features.

The interrelationship between homeostasis and steady state is well developed by Buckley who remarks:

> Only closed systems running down to their most probable states, that is, losing organization and available energy, can be profitably treated in equilibrium terms. Outside this context the concept of equilibrium would seem quite inappropriate and only deceptively helpful. On the other side, only open, tensionful, adaptive systems can elaborate and proliferate organization. Cannon coined the term "homeostasis" for biological systems to avoid the connotations of equilibrium, and to bring out the dynamic, processual, potential-maintaining properties of basically unstable physiological systems. In dealing with the sociocultural system, however, we need yet a new concept to express not only the structure-maintaining feature, but also the structure-elaborating and changing feature of the inherently unstable system. The notion of "steady state," now often used, approaches the meaning we seek if it is understood that the "state" that tends to remain "steady" is not to be identified with the particular structure of the system. That is, as we shall argue in a moment, in order to maintain a steady state the system may change its particular structure. . . . Thus, the complex, adaptive system as a continuing entity is not to be confused with the structure which that system may manifest at any given time. Making this distinction allows us to state a fundamental principle of open, adaptive systems: Persistence or continuity of an adaptive system may require, as a necessary condition, change in its structure, the degree of change being a complex function of the internal state of the system, the state of its relevant environment, and the nature of the interchange between the two.[2]

---

[1]Ross Stagner, "Homeostasis," International Encyclopedia of the Social Sciences, VI (1968), 499-502.

[2]Walter Buckley, "Society as a Complex Adaptive System," in Modern

The societal characteristics of system persistence and adaptation also can be found in the writings of Talcott Parsons, who, in his functional approach to the study of society, sets forth four functional imperatives that a society must meet in order to survive. These imperatives are: (1) pattern maintenance, (2) adaptation, (3) goal attainment, and (4) integration.[1] It is the first two imperatives with which we are most concerned in this study, although the latter two are considered from time to time. On the one hand, there are strong tendencies to perpetuate the existing structure of authorities and areal political forms as well as certain beliefs and attitudes; but, on the other hand, there are also strong pressures at times for change and reform. In reality, the political system adopts compromises between these two imperatives; and these compromises will now receive our attention.

The tendency towards system persistence or pattern maintenance of existing areal political forms is deeply rooted in society. Once an areal political form has been established, it becomes part of an institutional framework protected by law, managed by a power structure, and staffed by a bureaucracy. These elements tend to discourage innovation and inhibit the ability of the political system to respond to new conditions. Moreover, cultural values and traditions are acquired by people living in the area and are reinforced by common usage. The political area molds the activity fields[2] of political and social groups in which support for the political areal form is generated. Not only do these factors inhibit change, but they also influence the milieu within which subsequent decisions are strongly conditioned. Through time a set of attitudes, loyalties, precedents, and laws tend to preserve the existing patterns.

Adaptation is the other societal functional imperative of importance in the creation of areal political forms. The environment that surrounds the political system is continually changing and creating new demands, but many aspects of the environment also exercise considerable constraints upon the freedom of action of authorities to respond to demands. The demands may be ignored or

Systems Research for the Behavioral Scientist: A Sourcebook, ed. by Walter Buckley (Chicago: Aldine Publishing Co., 1968), p. 493.

[1]Karl W. Deutsch, "Integration and the Social System: Implications of Functional Analysis," in The Integration of Political Communities, ed. by Philip E. Jacob and James V. Toscano (Philadelphia and New York: J. B. Lippincott Co., 1964), pp. 181-83.

[2]Activity fields are the spatial expressions of the system of man's interaction with his total environment. For an expanded discussion see: W. A. Douglas Jackson and Marwyn S. Samuels (eds.), Politics and Geographic Relationships: Toward a New Focus (2nd ed.; Englewood Cliffs, N.J.: Prentice-Hall, Inc., 1971), p. 6.

shunted aside for a time, but ultimately, the authorities must respond in a satisfactory manner or new authorities will be selected.

In response to demands from the environment, authorities have several options in order to maintain the steady state. One set of responses would be to modify the existing governmental structure. New officials could be selected, existing governments could perform additional services, or the boundaries of existing areal political forms could be extended. Another response would be to create entirely new governments, either single-purpose or multi-purpose.

The choices to be made are not always clear-cut, and much controversy results as reformist elements and political groups advocate one choice or another. In this study, we are primarily concerned with those choices that lead to the creation of Special Districts; therefore, our attention now will be focused on those aspects of the environment and political system that make Special Districts such desirable choices.

A suitable response to demands from the environment may be disruptive to the existing social order or political system, as it may be difficult to administer, conflict with the boundaries of existing areal political forms, require extensive changes in the attitudes of officials or citizens, or may involve extensive changes in the legal system. In many cases, the easiest way to restore the steady state is to create a Special District to solve the immediate problem, bypass reform, and leave the existing governmental structure intact. The activity field of the Special District is merely superimposed upon the activity fields of existing governments. Special Districts, then, can be considered as homeostatic mechanisms, enabling the political system to survive with minimal disturbance.

Using the concept of homeostasis, we can see how a pattern of areal political forms developed over time. Instead of being an optimal adjustment in which areal political forms are created and dissolved in response to some idealized value system, the pattern at any point in time reflects the accumulated and sometimes altered homeostatic adjustments that occurred before. As each new form was created, the political system adapted itself to the new structure and the steady state was restored. In turn, the areal political form, acting as a small sub-system, modified its behavior and goals through feedback so that it would continue to command support and be able to satisfy demands. This continuous renewal of raison d'être of an areal political form is an important point to consider in viewing the present framework because the environmental conditions that initially caused the formation of an areal political form may no longer be present.

Systems analysis enables us to look at the interaction of the political system and its encompassing environment in a multitude of situations and to identify common properties. With it as background, we now are ready to examine the individual constraints that have led to the formation of Special Districts in the United States.

Constraints upon the Political System Leading to the Creation of Special Districts

Numerous constraints, operating singly and in combination, cause the political system to create Special Districts rather than to modify the existing governments to fulfill demands. Changes in the environment cause citizens to present new demands to the authorities of the political system. At the same time, however, elements in the environment or in the authorities section itself may exercise so many constraints upon the system that the existing governmental forms cannot or will not fulfill all the needs of the citizens--a condition often referred to as "output failure." In these situations Special Districts are often very effective homeostatic mechanisms to restore the steady state. The Special Districts satisfy the immediate demands of the citizens and still do not materially affect the structure of existing governmental forms or the existing authorities.

The Advisory Commission on Intergovernmental Relations has studied many Special Districts throughout the country and has provided an excellent summary of the constraints acting upon the political system resulting in the creation of Special Districts.[1] The first eight constraints have been adapted largely from its work; the last two constraints have been added by the writer. These ten constraints are: (1) financial limitations; (2) limitations on the powers of local governments; (3) areal convenience; (4) political compromises; (5) desire for business management; (6) public acceptance of Special Districts; (7) programs of higher levels of government; (8) influence of special-interest groups; (9) the desire for independence; and (10) historical circumstances.

Financial Limitations

Debt and tax limitations on local governments imposed by State constitutions and by State statutes are often responsible for the creation of Special Districts. From time to time, local governments need to make improvements and

---

[1] Advisory Commission on Intergovernmental Relations, The Problem of Special Districts in American Government, pp. 53-63.

construct new facilities that require large capital expenditures. However, many units of local government are not in a position to finance the construction of large capital projects within current revenues nor are they able to borrow the necessary funds because constitutional and statutory limitations on debt and taxes prevent a local government from becoming indebted or levying taxes above a fixed percentage of assessed valuation. In order to circumvent these restrictions, Special Districts are often created with separate borrowing and taxing powers within the same area as the original unit of local government. This is legally possible because the restrictions are placed on the unit of local government, not on the area occupied by the unit of government. Thus a unit of area theoretically can have unlimited debt or taxes merely by creating additional layers of government.

The constitutional debt limitation is not as restrictive today as it once was because of three considerations. First, an increasing number of State legislatures have authorized local governments to acquire indebtedness outside constitutional and statutory debt limits for revenue-producing operations. Second, the constitutional debt limits themselves may have been made less restrictive by a new constitution. Third, the continued increase in property values, accompanied by efforts to more closely relate assessed valuation to market value, has greatly increased the opportunity for the local government to incur debt and collect additional taxes.

Limitations on the Powers of Local Government

The legal system has imposed two significant limitations on the powers of local government that have led to the increase in the number of Special Districts. These limitations are: (1) strict construction of powers granted to local government, and (2) the inability of local governments to establish differential taxing areas within their boundaries.

Strict Construction of Powers.

--Under the strict construction of powers interpretation, the local government can only perform services specifically authorized by law. For services that are not spelled out in the State constitution or statutes, the local unit of general-purpose government may have to resort to the formation of a Special District in order to perform the service. This strict interpretation, known as "Dillon's rule," is described as follows:

It is a general and undisputed proposition of law that a municipal corporation possesses and can exercise the following powers, and no others: First, those granted (by the state) in express words; second, those necessarily or fairly implied in or incident to the powers expressly granted; third, those essential to the accomplishment of the declared objects and purposes of the corporation--not simply convenient, but indispensable. Any fair, reasonable, substantial doubt concerning the existence of power is resolved by the courts against the corporation, and the power is denied. [1]

Differential Taxing Areas. --Many local governments are legally required to apply taxes uniformly throughout their jurisdictions, even though the service may be required in only a part of the governmental area. Unless this limitation can be circumvented by a special assessment (not always legally possible), it is difficult for a unit of general-purpose local government to provide a service to a particular area within its jurisdiction, even though the government may have the legal authority to provide the service. The Special District permits the service to be offered to a particular area within a larger jurisdiction, and those residents living within the area pay an additional tax.

For example, residents of a small unincorporated community in a large county want fire protection service. The home sites have wells and septic tanks, as well as other utilities, and there is no need to incorporate into a municipality. However, the county is unable to provide fire protection because it does not have the legal authority or because it cannot establish subordinate taxing areas. In this situation, the residents of the community wanting fire protection may establish a fire protection district, thereby matching up the area in which the costs are borne by those receiving the benefits.

Still another example of this proscription against differential taxing areas leading to the formation of Special Districts occurs in the built-up areas lying adjacent to cities (commonly called fringe areas). In these areas which lack municipal government, the desire for urban services comes irregularly. Residents normally will want streets, sewage disposal, water supply, and fire protection at different periods in the evolution of the neighborhood. The differences in time in the articulation of these desires for services often discourages thoughts of annexation or separate municipal incorporation. Where the county or township is not equipped or not authorized to provide the service within the particular neighborhood, the neighborhood residents will often resort to a series of overlapping Special Districts to provide the services.

---

[1] Robert L. Lineberry and Ira Sharkansky, Urban Politics and Public Policy (New York: Harper & Row, 1970), pp. 113-14.

Areal Convenience

The limitations imposed by the existing boundaries of units of general-purpose government often lead to the formation of many Special Districts. The service areas of particular functions do not necessarily correspond with the political boundaries of existing governmental forms. This non-correspondence between political areas and functional areas is of four general types: (1) Natural features of the physical environment may dictate the territorial scope of the function; (2) certain types of urban services may extend beyond municipal limits; (3) numerous units of general-purpose government may exist within a single service area of a particular function; and (4) functional areas may lie astride State boundaries.

In rural areas, where topographic features often dictate the area of a particular function, the Special District provides an expedient solution. For example, in watershed districts the stream and its entire catchment area may form a single district, yet lie in several counties. By having a single district, the politically undesirable act of combining the counties is avoided; yet the district may impose land-use controls and build structures for flood control.

Many people living outside municipal boundaries may desire urban services yet they do not wish to become part of the municipality. The district device permits a single unit of government to include the municipality and adjacent territory. In this way, the unincorporated area can have municipal services such as fire protection and parks; yet the residents of the unincorporated area can avoid the political control of authorities in the municipality.

People living in adjacent areas but under different jurisdictions of general-purpose government may share a particular service through the district device. Rather than facing the politically unpopular course of consolidating the existing governmental units, it is far simpler to create a Special District serving the different units. The individual units may maintain their independence in other functions.

Special Districts may be used to enable States to cooperate with each other in a function that is local in character but crosses a State boundary. The well-known New York Port Authority, for example, handles port facilities and certain forms of transportation in both New York and New Jersey. These interstate Special Districts are a mechanism by which States can maintain political control. If the district device were not available, the functions that span State boundaries would have to be assumed by the Federal government.

Political Compromises

Special Districts sometimes result from political compromises among conflicts of interest. In general, it is easier to create a Special District than it is to make changes in the form of general-purpose government. Such situations are common in unincorporated built-up areas near cities. The residents desire urban services; yet they do not want to change from a rural government (often a township) to a municipal government. It is easier to create a Special District to satisfy the immediate need than it is to bring about municipal incorporation or annexation.

In many areas, the unit of general-purpose government has the authority to provide a particular service yet does not wish to do so. The members of the power structure of the local government may not wish to provide additional services because they do not wish to risk citizen displeasure by raising taxes or they do not perceive the requested new services as sufficiently important. The creation of a Special District for the new service avoids a raise in taxes by the existing government and avoids a confrontation between the citizens who want the new service and the members of the power structure who do not want it.

At times Special Districts are able to include high tax-producing properties within their boundaries, an option not possible for the municipal government. An Oregon study gives an example of a park and recreation district which included the city of Springfield as well as the outlying areas of the community.[1] The civic groups in the city were extremely interested in creating a Special District because it would include industrial property outside the city limits. An earlier attempt to annex the industrial complex to the city had failed.

Desire for Business Management

Special Districts resulting from this factor are based on reasoning that follows from two implicit assumptions about the superior qualities of certain forms of societal organization. The first assumption is that actions influenced by business standards are superior to those influenced by governmental standards. The second assumption is that there ought to be a readily discernible connection between those who receive services and those who pay for them.

Many services now performed by government were once performed by private enterprise. This shift of services from the private sector to the public sector has not been necessarily accompanied by a shift in public attitudes as to

---

[1]University of Oregon, Problems of the Urban Fringe, II, 74-76.

what is efficient and what is not. The Special District is a convenient device by which the alleged superiority of private enterprise can be combined with the supposed social advantages of government.

In many Special Districts, particularly those financed through user charges, the service charges are easily determined and easy to justify. Once the judgment is made to finance a service through user charges, it is an easy step to believe that the service should be self-supporting. If the service is to be self-supporting, it is then argued that the service must be conducted in a businesslike manner. A logical continuation of such arguments is that responsibility for the function should be removed from politics. It is then argued that if this is not done, political influence will play too large a role in providing the service, and the service will not be provided efficiently.

Once the idea of a self-supporting district is accepted, and freedom from the general political system is achieved, it is easy to concentrate on efficient operations. Patronage can be bypassed, and the district can fill its positions with a higher caliber of personnel than are available to general-purpose government. There are also other advantages as noted by Alderfer, ". . . the functioning of the authority is not hampered by the detailed and often ridiculous statutory restrictions that bind regular municipal officials."[1] Along this same line, Tobin writes:

> The effective modern authority makes available the techniques of good business administration and management for public enterprises. It places them on their own feet and puts them on their own responsibility. It frees them from political interference, bureaucracy, and red tape without divorcing them from public control.[2]

Public Acceptance of Special Districts

Political leaders of a unit of general-purpose government often are not anxious to propose that their governmental unit assume the additional burden of providing a new service. Through experience, the leaders have found that the voters, faced with a referendum to incur debt, or to provide an additional tax levy for an expanded service, have often voted down the proposition. Yet, if the election were held to form a Special District to undertake the function or service, the vote often would be favorable. As noted in an Illinois study:

---

[1]Harold F. Alderfer, American Local Government and Administration (New York: Macmillan Co., 1956), p. 370.

[2]Austin J. Tobin, "Administering the Public Authority," Dun's Review (June, 1952), p. 72.

Whatever may be said for the desirability of making local government really local rather than local only in the sense that the A & P is local, it should be remembered that city and village government officials shy away from providing services that mean tax increases. Frequently the need for referendum approval of a new tax levy is the death knell of a critical service. Special District development has been something of a soporific in that a rejected levy for city park purposes creates no excitement when it appears on the tax bill as a rate for park district purposes. [1]

Public acceptance of Special Districts is particularly high where the service to be performed is financed through user charges. The creation of the district permits the service to be provided without it appearing to be a specific burden on the taxpayer. If the function is performed by a unit of general-purpose government, even if service charges defray the cost, the underlying support for the service comes from the property taxpayers. Presumably, the property owner feels that by voting against the service to be provided by the unit of general-purpose government, he avoids a basic liability.

When the service can be financed wholly or in part by user charges rather than from general tax revenues, the Special District has great appeal. This type of Special District, called a public authority by many modern writers, is very popular in certain States. The district provides a form of insurance for the property owner, in that his tax rate will not be affected by expenditures for the provision of services from which he does not benefit. The unit of general-purpose government is permitted to disregard the services provided by the district in its regular budget and in fixing the tax rates to finance it. Another advantage is that the operation of the service and its justification is removed from various State budgetary reviews.

Programs of Higher Levels of Government

Many of the Special Districts in the United States operate in fields in which the Federal government is active. Functions performed in urban areas by Special Districts for which Federal grant programs are available include libraries, hospitals, airports, parking, housing, sewage disposal, and mass transportation. Federal programs also affect many of the natural resource districts, including soil conservation, drainage, flood control, and irrigation.

The present-day impact of Federal programs must be considered in combination with other factors when speaking of the formation of Special Districts. Such factors include the lack of authority of general-purpose governments to

---

[1] Snider and Anderson, Local Taxing Units: Illinois Experience, p. 13.

assume certain types of responsibilities, to enter certain types of contracts, or to participate in joint undertakings.

For example, Federal funds are available for the construction of flood protection and water conservation facilities in small drainage basins. Such drainage basins are usually within a portion of a county or portions of more than one county. Unless the county is authorized to undertake the function, the availability of Federal grant funds will provide a good reason for the creation of Special Districts.

A common type of Special District in urban areas is the housing authority, coterminous with the municipality. The Federal Housing Act of 1937 supplanted public housing construction and operation by the Federal government with Federal financial assistance to local governments that were prepared to undertake housing projects. It was legally impossible for many municipalities to issue bonds, and there was some question as to whether municipalities could engage in such a function as housing. These objections could be bypassed by creating a local housing authority, and the prospect of Federal aid prompted the growth of many housing authorities. The original housing act was supplemented by many other housing acts, including slum clearance and redevelopment. The housing authority provided a convenient mechanism for direct negotiations between the Federal government and local governments, bypassing the State government.

Influence of Special-Interest Groups

For convenience in discussion, the special-interest groups will be divided into broad categories. The first group includes citizen groups which are concerned only with a particular function, with little regard for the over-all impact on government. The second group includes those individuals and enterprises which stand to benefit economically from the creation of a Special District.

Citizen Groups. --The citizen groups include public-spirited people and various organizations, which are interested in a particular aspect of governmental services. Citizens interested in public parks, for example, may find it easier to secure the necessary financial base to provide an adequate park system if a Special District is created than if the park system were provided by a unit of general-purpose government. The creation of a park district removes the service from the vicissitudes of the every-day policy-making processes of government. The unit of general-purpose government may be governed by officials who do not wish to provide the service or feel that there are more urgent needs for the available resources. During the early decades of the twentieth century, many park districts were created within the city limits of Chicago. Even though the city had the authority to provide parks, it fell far short of citizen demands for park facilities; and neighborhood groups created their own park districts to provide the facilities.

Special Districts have great appeal for citizen groups because of the way in which the districts function outside the political system of general-purpose governments. The groups are more concerned with particular services performed by government than the government process as a whole. If the groups find inadequacies in the performance of services in which they are interested, they often will resort to the device of Special Districts as this is a more feasible mechanism than trying to change the officials in general-purpose governments.

Once Special Districts are created, they develop constituencies that strongly resist change, even though the officials of general-purpose governments have changed and may be willing to provide the services. As Bollens remarks, "People and groups possessing a major interest in one function frequently resist having the function allocated to an established general-purpose government or even another special district."[1]

Economic Benefactors. --The second type of special-interest groups influencing the use of Special Districts consists of various individuals and enterprises which stand to benefit economically from the creation of a district and its continued existence. This group includes attorneys, sellers of bonds, financial advisors, bank trustees, engineers, and public accountants. Bollens is very critical of some of these special-interest groups. As he puts it:

> Some actions in favor of special districts, however, are so baldly based on complete selfishness that they warrant brief separate consideration. One illustration relates to the actions of private concerns anxious to sell equipment and supplies. Judging that their business opportunities will be enhanced, they sometimes provide the principal stimulus for the establishment of special districts. The organization of supporting "citizen groups," the payment of election fees, and the circulation of district formation petitions and nominating papers of sympathetic governing body candidates are all techniques that have been employed by businesses acting wholly in their own self-interest. The result in one instance was the creation of a sanitary district which laid sewer pipes far in excess of the needs of both the present and the foreseeable future population. Another example of self-interest

---

[1] Bollens, Special District Governments, p. 10.

is the desire of local residents to realize a return on the tax money collected in their area. This self-centered attitude explains the establishment of a number of road districts in Missouri. Another far from altruistic reason for creating road districts is that they provide employment opportunities, in construction and maintenance, for governing body members and their relatives and friends. [1]

Another more indirect benefit to a few individuals concerns private real-estate developers which may benefit from the creation of a Special District to finance construction of sewer lines or water mains. Such districts permit the cost of the improvements to be spread out over the life of the bonds issued for construction of the improvements rather than appearing as a factor in the actual selling price of the house. In effect, the buyer pays a higher price through the technique of user charges and taxes, both of which are used to retire the bonds.

Desire for Independence

The desire for independence and autonomy is often an important factor in the creation of Special Districts. In many localities, admiration is expressed for a government that is small in area or small in the number of functions performed. Instead of making an existing government more complex through functional enlargement, a new more simplified government is created. This action is not only related to a primary interest in a single function, but it is also a manifestation of the belief that government can be better observed and controlled, if it is kept small. [2]

Echoing this same theme, Asseff writes of Special Districts in Louisiana:

. . . the special district is used because it allows considerable local control. As a result, there is a close relationship between administration and political responsibility, in that responsibility is for a single function of government and the officials of the district are ordinarily either popularly elected by the citizens of the area or are appointed by some local authority. [3]

Capitalizing on this distrust of large general-purpose government, many officials of Special Districts wish to remain outside the purview of general-purpose government. As a separate entity, they feel they can utilize available resources more effectively, can be more responsive to citizen demands, can hire their own personnel, and can remain free from the political domination of general-purpose government.

As an example of this desire for autonomy, we may note the formation of a library district in a suburb northwest of Chicago. The municipal library was

---

[1] Ibid., pp. 14-15.   [2] Ibid., pp. 10-11.

[3] Asseff, Special Districts in Louisiana, p. 3.

not adequately serving the citizens of the municipality, and the municipal officials would not permit sufficient funds to be spent to improve the library service. Taking advantage of Illinois law, an interest group concerned with the
improvement of library service succeeded in having a library district established that had the same boundaries as the municipality. The library district
now had its own tax base and was able to greatly improve the quality of library
service.[1]

Historical Circumstances

In a few cases, Special Districts are the modern legal forms of past
types of organizations. These organizations performed a specific function, and
the advantages of these organizations enabled them to be perpetuated in modern
laws. Two examples of this type of district will be given: fire protection districts and community land-grant districts.

Historically, fire protection was provided largely by volunteer companies. When it became necessary to establish a more efficient service, these
companies wanted to maintain the community and social activities associated
with these groups, amenities which would be lost if fire protection was provided
by general-purpose governments. By establishing fire protection districts, the
desirability of continued group interaction could be preserved, and the districts
could upgrade their fire protection to the level they wished.[2]

The community land-grant districts in New Mexico are a carry-over of a
structure which existed under the Spanish occupation of the territory from 1693
to 1821. The early Spanish farmers, who settled along streams on privately
owned lands, had their grazing lands in common ownership in the form of land
grants. These forms were perpetuated by the Mexican government from 1821
until 1846 when the lands came under American military rule. The legal history of these forms from 1846 to the present time is much too complex to relate
here, but in any case, these forms were recognized by the Census Bureau as
Special Districts in 1957.[3]

---

[1] Interview with Chief Librarian of the Franklin Park Library District,
August 7, 1974.

[2] Alderfer, American Local Government, p. 530.

[3] Folmar, Special Districts in New Mexico, p. 26.

Conclusions

This chapter has shown how the political system utilizes Special Districts to provide services that cannot or will not be furnished by general-purpose governments. Many factors in the environment and the political system itself cause this output failure of existing general-purpose governments, but, at least in the short run, Special Districts appear to be the most expeditious way for the political system to achieve homeostasis.

The factors underlying the formation of Special Districts combine with one another in many ways, and the particular set of factors that combine to form a particular district varies from place to place. In the following chapter, we shall deal with the structure of governmental forms of the local political system in a particular area--Cook County, Illinois. Once this structure has been outlined, we shall then be in a position to examine the individual types of districts. As each type of district is examined, we can then view the interaction of the political system of Cook County with the particular set of factors in the environment that led to the formation of the district type.

CHAPTER III

GOVERNMENTAL FRAMEWORK IN COOK COUNTY

In this chapter the discussion shifts from the general topic of Special Districts as governmental forms to the case of Special Districts in Cook County. In a case study, it is possible to integrate the Special District development with the particular combination of environmental sub-systems and historical factors that exist in any area. The chapter begins with a brief introduction to the governmental forms in Cook County. With this introduction as background, there follows a historical account of Special District development in the county. The chapter concludes with the selection of Special Districts to be further studied and the organization of the succeeding chapters.

The State of Illinois has authorized four types of local government in Cook County. These types are the: (1) county; (2) township; (3) municipality; and (4) Special District. Each of these types is empowered by the State to perform certain functions, but a particular function need not necessarily be performed by only one type of government. In some instances, several types of government perform the same function; in other words, the functions overlap. The county, township, and municipality are general-purpose governments and are empowered to perform many different functions within their jurisdictions. The Special District, on the other hand, is a single-purpose government, generally confined by law to but a single function.

Cook County Government

County organization has its roots deep in English history, which accounted for its early development in the American colonies. The county was perpetuated in the Northwest Territories, and county government already existed in Illinois when statehood was achieved in 1818.[1]

---

[1] League of Women Voters of Illinois, Structure of Local Government in Illinois: The County (Chicago, 1968), p. 1.

Cook County was organized in 1831. After other counties were formed from part of its territory, its present boundary was established in 1839.[1] Since that time other governments and institutional structures have been greatly influenced by this boundary, even though technological and cultural change have almost completely altered the physical and social environment within which the boundary originally was drawn.

From 1831 to 1850, Cook County was governed by a three-man Board of Commissioners. From 1850 to 1871 Cook County was governed by township supervisors, a form of government still used by all other Illinois counties that have township organization.[2] Due to the rapid growth of Chicago and its continual under-representation on the governing board of Cook County, the township-supervisor form of government was replaced in 1871 by a Board of Commissioners, a governmental form which still operates today.[3]

The Cook County Board is comprised of fifteen commissioners, ten of whom must come from Chicago and five of whom must come from the remainder of Cook County. This composition was prescribed by State law over a century ago, even though boundary changes and population shifts have made the legal ratio somewhat unrealistic.[4] In the 1970 census, Chicago had about three-fifths of the Cook County population, which is slightly less than the two-thirds that would be required to comply with the legal ratio.

The Cook County government has a variety of functions. Some of these functions are performed exclusively by the county, but many of them are shared with other units of government. The chief county functions are law enforcement, maintaining roads and bridges, assessment of property, collection of property taxes, conducting elections, recording legal documents, provision of aid for the aged and infirm, protection of health, and supervision of education. In performing these functions, it maintains a jail, hospital, and home for the aged. The

---

[1] Cook County League of Women Voters, The Key to Our Local Government (Chicago: Citizens Information Service, 1966), p. 64.

[2] Of the 102 counties in Illinois, eighty-five have townships and seventeen do not. Cook County is the only county with townships that also has commissioners. The other eighty-four counties with townships have the township-commissioner form of government. The seventeen non-township counties have three commissioners elected at large.

[3] Charles B. Johnson, Growth of Cook County, I (Chicago: Board of Commissioners of Cook County, 1960), 91-92.

[4] Laws of Illinois (1871), p. 308.

county also exercises the regulatory powers of zoning and licensing in unincorporated areas.[1]

Township Government

Illinois originally was settled by people who came from Kentucky and Tennessee where there was no township organization. By the 1830's, however, the opening of the Erie Canal and removal of the Indian threat had brought many settlers from the northeastern states where township organization was strong.[2] As a compromise between these two groups of settlers with different heritages, the Illinois Constitution of 1848 permitted each county to decide for itself whether it wished township organization.[3] In November, 1849 the Cook County voters adopted the township form of organization, and twenty-seven of the present thirty-eight Cook County townships were organized in April, 1850.[4]

Although townships were originally rural governmental forms, numerous attempts were made to adapt township government to an area which fast was becoming urbanized. In the latter half of the nineteenth century and early part of the twentieth century, several township boundaries were altered, and in the more densely populated areas eleven new townships were created from parts of the older ones. Figure 2 shows the present townships in Cook County. Even as new townships were being formed, the older township organization in Chicago became unworkable. In 1901 township organization was abolished in Chicago, and the township duties were transferred to the county or assumed by the city.[5] The township areas, however, still are used inside Chicago by county tax officials. Within Chicago, there are eight whole townships and parts of nine others.[6] In those cases where the Chicago boundary extends across a township, the part of the township outside the city has a separate legal government; the part within the city retains its areal form but has no legal existence.

---

[1] Summarized from Cook County League of Women Voters, Key to Local Government, pp. 67-131, passim.

[2] Johnson, Growth of Cook County, pp. 91-92.

[3] Illinois, Constitution (1848), art. vii, sec. 6.

[4] Johnson, Growth of Cook County, p. 92.

[5] Charles E. Merriam, Report of an Investigation of the Municipal Revenues of Chicago (Chicago: City Club of Chicago, 1906), p. 12.

[6] In 1958 the municipal boundary of Chicago was extended into Addison Township of Du Page County in order to place all of O'Hare Airfield within the city.

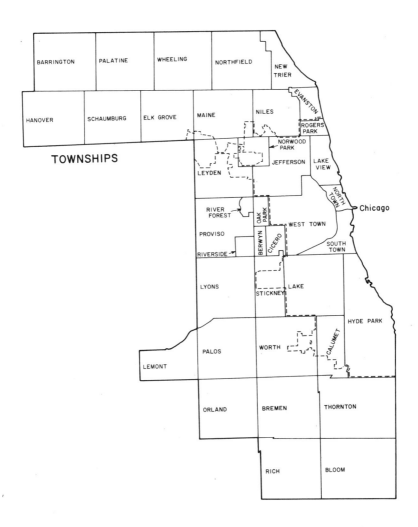

Fig. 2.--Townships in Cook County

For many decades, townships in Cook County have been slowly losing their relative importance as other units of government have assumed new func- tions. The State Legislature has been very reluctant to give townships addi- tional powers so that they could be responsive to new citizen demands. Despite this decline in importance, the county political organizations are organized by townships outside Chicago, a circumstance that probably assures the survival of the townships. The chief functions of the townships today are poor relief, assessment of personal property, and maintenance of roads in unincorporated areas.[1] In five cases, municipalities are territorially coincident with town- ships (Berwyn, Cicero, Evanston, Oak Park, and River Forest); in these areas the township has no road function.

Municipal Government[2]

Legal History. --Municipal government goes back to ancient times and has been present in the United States since the colonial period. In the early his- tory of Illinois, municipalities were established by special acts of the State Legislature. The Legislature maintained close control over municipalities, and many local matters were handled directly by the Legislature.

With rapid urbanization in the latter half of the nineteenth century, the existing governmental framework in Cook County experienced great stress in adjusting to the needs of a society that was increasing rapidly both in population and a need for services. Rather than increasing the powers of the county and its townships to cope with increased demands, the Legislature decided that the urban services could best be provided by other governmental forms. The most widely used forms authorized by the Legislature were municipalities, and to a smaller extent, Special Districts.

In 1872 the Legislature permitted the formation of cities and villages under a general enabling act.[3] This act spelled out the powers of municipalities and facilitated the formation of many new municipalities. It also enabled munic- ipalities to have more freedom of action and relieved the Legislature from hav-

---

[1] League of Women Voters of Illinois, Structure of Local Government in Illinois: The Township (Chicago, 1968), p. 4.

[2] With the exception of Cicero, all municipalities in Cook County are vil- lages or cities. Cicero is a town created by the Legislature in 1867. This type of municipal government no longer is authorized by the Legislature.

[3] Laws of Illinois (1872), pp. 218-73.

ing to deal with many purely local matters. Except for ten old municipalities, most of the municipalities in Cook County were created under this act or its amendments (Table 3). In 1973 there were 128 municipalities in or partly in Cook County (Figure 3).

TABLE 3

DATE OF INCORPORATION OF MUNICIPALITIES
IN COOK COUNTY[a]

Date of Incorporation	Number Incorporated	Cumulative Total
1830-1839	1	1
1840-1849	1	2
1850-1859	2	4
1860-1869	6	10
1870-1879	5	15
1880-1889	8	23
1890-1899	24	47
1900-1909	14	61
1910-1919	16	77
1920-1929	18	95
1930-1939	1	96
1940-1949	8	104
1950-1959	20	124
1960-1969	3	127
1970-1973	1	128

Source: Compiled from Northeastern Illinois Planning Commission, Suburban Factbook (Chicago: Northeastern Illinois Planning Commission, 1971), pp. 48-54.

[a]Includes municipalities which were originally incorporated in other counties but have extended their boundaries into Cook County.

The major functions performed by municipalities are police and fire protection, water supply, sewerage systems, maintaining and cleaning streets, collecting garbage and trash, providing libraries, regulating traffic, and rodent control. Regulatory functions include zoning, licensing, inspection, and planning.[1] In some municipalities, these functions may be performed by other governmental forms or not performed at all.

---

[1]Summarized from Cook County League of Women Voters, Key to Local Government, pp. 5-61, passim.

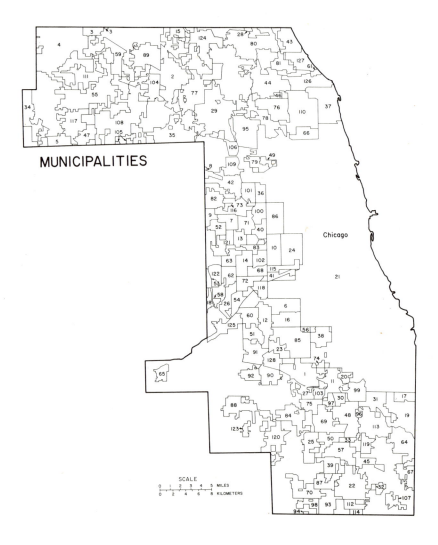

MUNICIPALITIES

Chicago

SCALE
0 1 2 3 4 5 MILES
0 2 4 6 8 KILOMETERS

Fig. 3.--Municipalities in Cook County

1. Alsip
2. Arlington Heights
3. Barrington
4. Barrington Hills
5. Bartlett
6. Bedford Park
7. Bellwood
8. Bensenville
9. Berkeley
10. Berwyn
11. Blue Island
12. Bridgeview
13. Broadview
14. Brookfield
15. Buffalo Grove
16. Burbank
17. Burnham
18. Burr Ridge
19. Calumet City
20. Calumet Park
21. Chicago
22. Chicago Heights
23. Chicago Ridge
24. Cicero
25. Country Club Hills
26. Countryside
27. Crestwood
28. Deerfield
29. Des Plaines
30. Dixmoor
31. Dolton
32. East Chicago Heights
33. East Hazelcrest
34. Elgin
35. Elk Grove Village
36. Elmwood Park
37. Evanston
38. Evergreen Park
39. Flossmoor
40. Forest Park
41. Forest View
42. Franklin Park
43. Glencoe
44. Glenview
45. Glenwood
46. Golf
47. Hanover Park
48. Harvey
49. Harwood Heights
50. Hazelcrest
51. Hickory Hills
52. Hillside
53. Hinsdale
54. Hodgkins
55. Hoffman Estates
56. Hometown
57. Homewood
58. Indian Head Park
59. Inverness
60. Justice
61. Kenilworth
62. La Grange
63. La Grange Park
64. Lansing
65. Lemont
66. Lincolnwood
67. Lynwood
68. Lyons
69. Markham
70. Matteson
71. Maywood
72. McCook
73. Melrose Park
74. Merrionette Park
75. Midlothian
76. Morton Grove
77. Mount Prospect
78. Niles
79. Norridge
80. Northbrook
81. Northfield
82. Northlake
83. North Riverside
84. Oak Forest
85. Oak Lawn
86. Oak Park
87. Olympia Fields
88. Orland Park
89. Palatine
90. Palos Heights
91. Palos Hills
92. Palos Park
93. Park Forest
94. Park Forest South
95. Park Ridge
96. Phoenix
97. Posen
98. Richton Park
99. Riverdale
100. River Forest
101. River Grove
102. Riverside
103. Robbins
104. Rolling Meadows
105. Roselle
106. Rosemont
107. Sauk Village
108. Schaumburg
109. Schiller Park
110. Skokie
111. South Barrington
112. South Chicago Heights
113. South Holland
114. Steger
115. Stickney
116. Stone Park
117. Streamwood
118. Summit
119. Thornton
120. Tinley Park
121. Westchester
122. Western Springs
123. Westhaven
124. Wheeling
125. Willow Springs
126. Wilmette
127. Winnetka
128. Worth

Growth of Municipalities.--Only in the first decade of its existence did Cook County have a rural population that exceeded urban population (Table 4). The history of this urban population has been intertwined with the history of Chicago which was platted in 1830, organized as a town in 1833, and incorporated as a city in 1837.[1] Chicago had its greatest proportion of Cook County population (92.4 per cent) in 1900, although the city did not reach its maximum size until 1950. On the other hand, the population of municipalities outside Chicago has been increasing in absolute numbers and in percentage of Cook County population for nearly a century. The proportion of population in unincorporated areas has been a very small part of the total population since 1890.

TABLE 4

POPULATION OF COOK COUNTY, CHICAGO, NON-CHICAGO
INCORPORATED PLACES, AND UNINCORPORATED
TERRITORY, 1840-1970

Census Year	Cook County	Chicago	%	Non-Chicago Incorporated Places	%	Unincorporated Territory	%
1840	10,201	363	3.6	---[a]	-	9,838[b]	96.4
1850	43,385	29,963	69.1	---[a]	-	13,422[b]	30.9
1860	144,954	112,172	77.4	---[a]	-	32,782[b]	22.6
1870	349,966	298,977	85.4	---[a]	-	50,989[b]	14.6
1880	607,524	503,185	82.8	26,733	4.4	77,606	12.8
1890	1,191,922	1,099,850	92.3	47,729	4.0	44,343	3.7
1900	1,838,735	1,698,575	92.4	105,622	5.7	34,538	1.9
1910	2,405,233	2,185,283	90.9	176,671	7.3	43,279	1.8
1920	3,053,017	2,701,705	88.5	306,345	10.0	44,967	1.5
1930	3,982,123	3,376,438	84.8	567,522	14.2	38,163	1.0
1940	4,063,342	3,396,808	83.6	607,318	14.9	59,216	1.5
1950	4,508,792	3,620,962	80.3	781,338	17.3	106,492	2.4
1960	5,129,725	3,550,404	69.2	1,398,450	27.3	180,871	3.5
1970	5,492,369	3,366,957	61.3	1,982,831	36.1	142,575	2.6

[a]Incorporated places other than Chicago not reported separately before 1880.

[b]Includes all population outside Chicago.

Source: Compiled from decennial reports of the U.S. Bureau of the Census, Census of Population, 1840-1970.

[1]Johnson, Growth of Cook County, pp. 72-75.

Around 1890 a basic structural change took place in the pattern of municipal governments in Cook County. Prior to 1889 most of the urban population lived in Chicago, and the limits of the city were periodically extended by acts of the Legislature or resolutions of the Cook County Commissioners.[1] By an election in 1889 Chicago was consolidated with four townships (Hyde Park, Jefferson, Lake, and Lake View); at this time Chicago achieved about four-fifths of its present area (228 square miles). Since 1889 annexations to Chicago have been small, and the great growth of population between 1890 and 1930 took place largely in areas that had been annexed earlier.

After 1870 the non-rural population of Cook County began to be dispersed along the railroads that radiated outward from Chicago. At first, this population lived in areas that subsequently were annexed to Chicago. Later on, however, new municipalities began to be organized that were able to retain their independence. These new municipal governments marked the advent of polarization of political power between Chicago and its suburbs. The percentage of population living in municipalities outside Chicago rose from only 4 per cent in 1890 to 36 per cent in 1970.

With the coming of the automobile in the 1920's and a great increase in accessibility, the pattern of suburbanization of many large United States cities began to change from the stellate pattern extending along the railroads outward from the central city. In Cook County, as in other rapidly urbanizing areas, the interstitial areas between the prongs of settlement began to be filled, particularly in those areas adjacent to expressways. The additional population was absorbed both by the formation of new municipalities and the expansion of many older railroad-oriented municipalities. Outside Chicago and the older suburbs the rapid growth in population is still continuing.

Special District Government

Why Special Districts Were Formed. --A century of continuous population growth in Cook County, accompanied by the transformation of a rural society into an urban one, has produced a set of demands for additional services that could not or would not be met by the existing general-purpose governments. As a result of these demands not being met, Special Districts were created by the political system to bypass the output failure of general-purpose governments and

---

[1]Information obtained from "Map of Chicago: Showing Growth of the City by Annexations and Accretions," prepared by the City of Chicago, Department of Public Works, Bureau of Maps and Plats, 1970.

to provide the desired services. Citizen demands for services were made to
the authorities of existing political units within a matrix of many constraints,
severely limiting the choice of responses that authorities could make. For ease
in discussion, the principal constraints on the political system that led to district formation may be grouped into three broad categories: (1) a restrictive
legal structure; (2) beliefs and attitudes of citizens and officials; and (3) inappropriate areas of existing governments.

Foremost among the constraints were the legal limitations placed on
general-purpose governments dating from a nineteenth century rural society.
Constitutional limitations dating from 1870 included debt and tax restrictions, a
lack of authorized powers, as well as the necessity for establishing uniform taxing areas. These limitations evidently were adopted to prevent the abuse of
power by general-purpose governments, but the complications resulting from
these limitations were not foreseen. Rather than reforming the State Constitution, it was far simpler to pass legislation for the creation of Special Districts,
thereby enabling the political system in effect to circumvent the restrictions.
Another legal difficulty arose from the unwillingness of the Legislature to be-
stow additional powers on local units of general-purpose government, even if it
were constitutionally possible. Rather than liberalizing the restrictive laws,
the Legislature preferred to make piecemeal concessions to public demands for
additional services by the passage of enabling legislation for single-service Special Districts.

Another set of constraints stemmed from the desires and attitudes of the
people themselves. These included the desire to maintain local control, the
reluctance to delegate powers to higher levels of government, the disinclination
of many officials of general-purpose governments to provide desired services,
and the lack of a desire for municipal government in unincorporated areas that
desired urban services. Many of these attitudes are deep-seated and not only
derive from limited perceptions but also from the frontiersman's pioneering
spirit of independence and a not wholly unjustified lack of confidence in higher
levels of government.

The area flexibility of the district device has made Special Districts feasible and expedient political instruments to overcome the rigid boundaries of
existing political units. Service areas often do not coincide with the boundaries
of general-purpose government, and the district mechanism makes it possible
to set up realistic service areas. For example, areas surrounding a municipality can be included with the municipality by creating a Special District for a

certain service, but these areas can remain free of municipal government for other services. In other cases, combining two or more units of general-purpose government in a single-service Special District may achieve limited functional cooperation, and a sense of independence can still be maintained by the general-purpose governments in the provision of other services. In still other cases, the creation of a Special District permits a part of a municipality or township to be set aside for a particular service not desired by the whole area of general-purpose government.

Historical Development. --The development of Special Districts in Cook County has continued for more than a century. Throughout this period, municipalities and townships have been maintained as units of general-purpose government, although for certain functions Special Districts have been superimposed over the areas of these governmental units.

In the latter decades of the nineteenth century, urbanization took place largely within the municipal limits of Chicago. Park districts had been superimposed over the city and adjacent townships in 1869, and most of the city was covered by a sanitary district in 1889. Except for these two services, urban services were provided largely by the Chicago municipal government.

In the 1890's however, the character of Special Districts began to change. The Legislature began to pass enabling legislation by which citizens could circumvent the output failure of municipal and township governments. Local interest groups were able to bypass officials of general-purpose governments and petition the courts for the formation of Special Districts to provide desired services. Table 5 provides the date the Legislature first passed enabling legislation for each type of district, as well as the first time the district appeared in Cook County.

Beginning in 1896 small park districts began to be formed in Chicago and suburban areas. After the start of the twentieth century, the number of districts began to increase more rapidly, numbering twenty-eight by 1917 (Table 6). Although the whole county was covered by a forest preserve district in 1915, the great majority of these early districts were small park districts formed in built-up areas. The park districts provided better and more intensive recreation facilities than the municipal governments were able to provide. Moreover, the districts had locally elected officials, which provided opportunities for local control.[1] By 1930 the number of districts had more than doubled to fifty-eight.

---

[1]Officials of the three original park districts in Chicago were chosen by the governor of the State or by the Circuit Court.

TABLE 5

SPECIAL DISTRICTS FORMED BY
GENERAL ENABLING LAWS[a]

Type	Date of Enabling Legislation	Date of First District in Cook County
Park	1895[b]	1896
Forest Preserve	1913	1915
Public Health	1917	1920
Outlying Sanitary	1917	1922
River Conservancy	1925	1956
Mosquito Abatement	1927	1927
Fire Protection	1927	1940
Collector Sanitary	1936	1938
Tuberculosis Care	1937[c]	1949
Library	1943	1959
Street Light	1949	d
Mass Transit	1959	1967

[a]The three large park districts in Chicago were created by special legislation in 1869. The Chicago Sanitary District (later known as the Metropolitan Sanitary District of Greater Chicago) was created in 1889 by supposedly general legislation, although the law could only apply to one place.

[b]This is the first enabling legislation that applies to park districts in Cook County. In 1893 a law was passed for the formation of pleasure driveway and park districts which are not found in Cook County.

[c]Legislation for tuberculosis districts in municipalities was first passed in 1908, but these were not independent Special Districts.

[d]Data not available.

Most were still small park districts, but the passage of enabling legislation had permitted districts to be formed for public health, outlying sanitary facilities, and mosquito abatement.

During the 1930's the number of districts declined, owing to the extensive park district consolidation in Chicago. Also during this decade, fringe districts in unincorporated areas, but furnishing urban services, first appeared. A collector sanitary district was formed about 1938, and two years later, the first fire protection district was formed in Cook County.

During World War II the pattern of Special Districts remained stationary, but after the war, Special Districts grew dramatically in number and in area.

TABLE 6

NUMBER OF SPECIAL DISTRICTS IN COOK COUNTY, 1869-1973

Type	1869[a]	1889[b]	1917[c]	1930[d]	1939[e]	1962[f]	1973[g]
Park	3	3	26	50	37	59	93
Metropolitan Sanitary (1889 type)	0	1	1	1	1	1	1
Forest Preserve	0	0	1	1	1	1	1
Public Health	0	0	0	1	1	2	2
Outlying Sanitary (1917 type)	0	0	0	3	4	3	2
Mosquito Abatement	0	0	0	2	2	4	4
Collector Sanitary (1936 type)	0	0	0	0	2	8	25
Fire Protection	0	0	0	0	0	39	45
River Conservancy	0	0	0	0	0	1	1
Tuberculosis Care	0	0	0	0	0	1	1
Library	0	0	0	0	0	2	14
Street Light	0	0	0	0	0	1	1
Mass Transit	0	0	0	0	0	0	6
Total	3	4	28	58	48	122	196

See Appendix II for sources and notes.

There was a sudden growth of population in the outlying parts of the county, with concomitant increased demands for services that far exceeded the capability of general-purpose governments to provide them. Districts proliferated in both incorporated and unincorporated areas as the pre-war dichotomy between rural and urban began to break down. Fringe districts, particularly collector sanitary and fire protection districts became widespread. Although previously serving municipalities, park districts became common in erstwhile rural areas with little regard for municipal boundaries.

In addition to increasing the number of Special Districts of previously existing types in the post-war decades, new districts were created for libraries, street lighting, river conservancy, tuberculosis control, and mass transit. By 1962 the total of 122 districts more than doubled the 1939 figure. After 1962 the number of districts continued to increase, reaching 196 in 1973. Most districts were formed in the newly urbanized parts of Cook County, although mass transit districts for many older municipalities made their initial appearance in 1967.

In Table 7 we see the growth in Special Districts by area, a statistic

TABLE 7

AREA (SQ. MI.) OF SPECIAL DISTRICTS IN COOK COUNTY, 1869-1973

Type	1869[a]	1889[b]	1917[c]	1939[d]	1973[e]
Park	136	146	211	343	659
Metropolitan Sanitary (1889 type)	0	182	375	439	864
Forest Preserve	0	0	959	959	959
Public Health	0	0	0	4	16
Outlying Sanitary (1917 type)	0	0	0	19	15
Mosquito Abatement	0	0	0	144	753
Collector Sanitary (1936 type)	0	0	0	2	21
Fire Protection	0	0	0	0	289
River Conservancy	0	0	0	0	2
Tuberculosis Care	0	0	0	0	733
Library	0	0	0	0	105
Street Light	0	0	0	0	0[f]
Mass Transit	0	0	0	0	140
Total	136	328	1,545	1,910	4,556

See Appendix II for sources and notes.

even more dramatic then the growth in number. The early districts were confined mostly to built-up areas, but in 1915 the entire county was blanketed by a single district. By 1939 the area of Special Districts had grown to 1,910 square miles for an average of two layers of Special District government for the entire county. After World War II the pattern of districts began to change, with districts forming in rural and built-up areas alike. By 1973 the area of Special Districts had increased to 4,556 square miles covering the county with an average of 4.8 layers of Special District government. At present, all parts of the county have at least two layers of Special District government, and some parts have as many as seven.

Selection of Special Districts for Further Study

In an area as complex as Cook County, the decision as to what is and what is not a Special District is not as easy as the preceding historical summary may suggest. The sources of information vary greatly in quality and at times are contradictory or ambiguous, necessitating numerous practical compromises between what is desirable and what is possible.

The Special Districts selected for further study will be the 196 taxing districts of Cook County.[1] Information for taxing districts is the most readily available and the most consistent over a period of time.

Comparability with Census Bureau Materials. --Fortunately, most governments in Cook County with the word district in their titles are independent governments and consequently are counted by the Census Bureau. Beyond this easy generalization, it is essential to be more exact and examine the districts in more detail.

Only in a very rough manner does the number of taxing districts obtained by map inspection approximate the Census Bureau figures in 1967. At that time, the Census Bureau listed 167 Special Districts in Cook County, of which 141 were taxing districts and 26 were not.[2] Since the Census Bureau did not identify the 167 districts it gave as a total, no further comparison with the 196 figure is possible. The 1962 Census of Governments, however, listed the districts by name.[3] Of the 158 districts listed by the Census Bureau in 1962, 122 were identified as taxing districts and 36 as non-taxing districts.[4]

The percentage of Cook County Special Districts that have the power of taxation (84 per cent in 1967 and 77 per cent in 1962) is unusually high when compared to the entire United States. For the whole country, about one-half the Special Districts have taxing power; for the State of Illinois, about three-fifths of the districts have such power.[5]

The question might well be asked, "What about the twenty-six governments that the Census Bureau considered as Special Districts in 1967 but are not taxing districts and are not shown on the maps of the Illinois Department of

---

[1]Taxing district means the district has the power to levy a tax; this power may or may not be exercised. The total figure of 196 districts is derived from maps prepared by the Illinois Department of Local Government Affairs. If any part of a district is in Cook County, it is counted, even though the larger part of the district territory may lie in an adjacent county.

[2]U.S. Bureau of the Census, Census of Governments: 1967, Vol. V: Local Government in Metropolitan Areas, p. 60.

[3]U.S. Bureau of the Census, Census of Governments: 1962, Vol. V: Local Government in Metropolitan Areas, pp. 507-8.

[4]Ibid.

[5]U.S. Bureau of the Census, Census of Governments: 1967, I, 72, 76.

Local Government Affairs?" Three of these districts are governments with unique functions: the Chicago Regional Port District, the Chicago Transit Authority, and the Metropolitan Fair and Exposition Authority. The remaining twenty-three districts are mainly concerned with housing, drainage, and water supply.[1] All of these non-taxing districts support themselves by user charges.

Of the taxing districts in Cook County not recognized by the Census Bureau, there are only three--the Forest Preserve District of Cook County, the Berwyn Public Health District, and the Stickney Public Health District. The Census Bureau considers these districts as dependent agencies of general-purpose governments and, therefore, do not meet the criterion of independence. Nevertheless, they are important in Cook County and are included in the ensuing discussion.

School Districts. --Although school districts are a type of Special District (in the old sense of the term), they commonly are excluded from the discussion of Special Districts in the current literature. The 1970 Illinois Constitution uses the term Special District, but it refers to school districts as distinct from Special Districts.[2] In keeping with current usage, the estimated 150 school districts of Cook County will not be included in the study.

Drainage Districts. --The estimated thirty-two drainage districts in Cook County are particularly troublesome objects of study owing to a lack of records.[3] Attempts to contact the officials of these districts proved fruitless, suggesting that the districts may exist legally but are not operating. In a 1933 study of government in the Chicago region, the authors had this to say:

> Because the true story of these governmental relics cannot be told without an examination of county, township, and even justice-court records, many of which have long been lost, it is highly probable there are several drainage districts in the Region which have not been located, and it is also probable that several of the districts counted have long ceased to operate. The lack of systematic records upon which even a system of unofficial co-operation might be based and the fact the Region exists as a coherent unit and is gradually supplanting the farms of the suburban area make the continued

---

[1] U.S. Bureau of the Census, Census of Governments: 1967, V, 60.

[2] Illinois, Constitution (1970), art. vii, sec. 1.

[3] The drainage districts are shown on a map, "General Drainage Map: Cook County," prepared by the Illinois Department of Public Works and Buildings, Division of Waterways, 1958.

existence of these districts in their independent and unrelated form an anomaly of metropolitan government. [1]

If anything, the situation regarding drainage districts has become worse in the past forty years, and these districts are excluded from further study.

Non-functioning Districts. --An occasional difficulty arises from the fact that not all taxing districts are necessarily active. Of the districts shown on the maps of the Illinois Department of Local Government Affairs, most are active as substantiated by personal investigation and by a check with the taxes levied by the County Clerk's office in the Cook County government. [2]

Some of the difficulty regarding non-functioning Special Districts is probably due to the legal complexities involved. As Bollens says:

> In numerous situations it is currently a legal impossibility to abolish certain types of districts, simply because no procedure for eliminating them has ever been enacted. As an illustration, there is no abolition process for many kinds of districts in Illinois, including fire protection, sanitary, hospital, tuberculosis sanitarium, street lighting, and water. Many districts become inactive without formally dissolving under an authorized procedure or seeking to obtain such a procedure when it does not exist. Numerous school districts as well as other governmental units, such as cities, have become inoperative. This practice can lead to confusion and to mistaken impressions by the casual observer. The governmental landscape is cluttered with many ghosts. [3]

The Cook County Clerk is supposed to notify the Illinois Department of Local Government Affairs of any change in district status or boundaries. In several cases this procedure apparently has not been followed, and considerable confusion results when field checks are made. Since the County Clerk's office does not permit inspection of its maps by the public, one can only speculate that taxes are not being levied for inactive districts or in areas that have been disconnected from existing districts. The maps in later chapters have been updated to reflect known changes.

---

[1] Charles E. Merriam, Spencer D. Parratt, and Albert Lepawsky, The Government of the Metropolitan Region of Chicago (Chicago: University of Chicago Press, 1933), p. 61.

[2] Cook County, County Clerk, "Comparative Statement of the Tax Rates for the Years 1969 and 1970."

[3] Bollens, Special District Governments, p. 20. In a few cases, legislation in Illinois now permits the dissolution of some of these districts. Although a fire protection district cannot be dissolved on its own initiative, a municipality that contains more than 50 per cent of the district may take over the district. Source: Illinois, Annotated Statutes, ch. 127-1/2, sec. 38.4. A sanitary district, organized under 1936 legislation, may be dissolved on the initiative of the district. Source: Illinois, Annotated Statutes, ch. 42, sec. 444.

Final Selection of Special Districts. --After the exclusion of school districts, drainage districts, and inoperative districts, we are ready for the final selection of districts for examination. The taxing districts were enumerated simply by tabulating them from maps. The type and number of these districts are as follows: parks (93), fire protection (45), sanitation (28), library (14), mass transit (6), mosquito abatement (4), public health (2), forest preserve (1), tuberculosis care (1), street lighting (1), and river conservancy (1).

Historical Data. --The historical data for Special Districts are very difficult to obtain. Inquiries at the Cook County Government have been singularly unproductive in regard to the records of Special Districts. Despite the lack of official data, fragmentary sources and personal interviews have made it possible to piece together a general picture of district development. The past areal extent of Special Districts must be reconstructed from a variety of sources. It was not until 1939 that all the taxing Special Districts were mapped systematically.

Organization of Chapters on Special Districts

Special Districts are extremely varied in their functions, legal authorization, and historical development. A simple taxonomy based on any of these characteristics is not possible. As a compromise, it was decided to group them by their relationships to other governmental forms. Such a grouping permits the functional utility of Special Districts to be brought out, since it was the output failure of other governments that led to the formation of so many Special Districts.

Chapter IV deals with the area-wide districts. These districts provide a few functions for large areas, representing a degree of intergovernmental cooperation not found in other districts. For the most part, district services are rendered with little regard for municipal or township boundaries.

Chapter V deals with districts that operate largely within municipalities. These districts perform urban services but are not part of the municipal government. Many of these districts developed because municipal governments were unable or unwilling to provide the services required. Some districts also permit municipal cooperation in a few services, although municipalities retain their independence by controlling other services. In a few cases, the district mechanism permits a part of the municipality to have a service which is not found in the rest of the municipality.

Chapter VI deals with districts in areas that are urban in character but for the most part lack municipal government. Since these districts usually lie adjacent to municipalities, they are termed fringe districts. In some cases, erstwhile fringe areas have acquired municipal government, but for certain services the district form of government is retained.

Chapter VII deals with Special Districts in summary form and discusses their impact on the space-polity of Cook County. The impact is discussed within the context of five aspects that possess spatial significance: (1) the size of Special Districts; (2) the distribution of resources; (3) horizontal integration; (4) centralization and local control; and (5) an appraisal of Special Districts.

CHAPTER IV

AREA-WIDE DISTRICTS

Introduction

The area-wide districts were created to provide certain services that units of general-purpose government could not or would not provide. Such inaction resulted from complexities of the legal system, inappropriate areas of general-purpose government, and a lack of commitment on the part of leadership of general-purpose governments. Rather than reforming the general-purpose governments, it was easier to create Special Districts as homeostatic mechanisms to perform the desired services and to leave the existing political structure relatively undisturbed.

Area-wide districts in Cook County cover extensive areas and bring about a degree of unification among municipalities and townships not found in other Special Districts. The functional cooperation among a host of jurisdictions through the mechanism of the area-wide district affords economies of scale, provides a large taxing base, and makes planning feasible for large areas. The quality of services given by these districts is remarkably uniform and reflects the egalitarian sentiments that are held in relation to the services provided by these districts.

The relatively few functions performed by these districts indicates the unwillingness of Cook County citizens to sacrifice local control and give up functions to large areal units of government. Fortunately, the services are of such a nature so as not to be desired by general-purpose governments, and the districts have been able to command wide support from the Cook County population.

The following table shows the area-wide districts in Cook County, together with their area and population (Table 8). Since the districts are so varied in their function, history, and legal status, they will be discussed individually in subsequent parts of the chapter.

58

TABLE 8

AREA-WIDE DISTRICTS IN COOK COUNTY

Name	Area		Population[a]	
	Sq. Mi.	Per Cent of Cook County	No.	Per Cent of Cook County
Forest Preserve District of Cook County	959	100	5,492,400	100
Metropolitan Sanitary District of Greater Chicago	864	90	5,328,000	97
Suburban Cook County Tuberculosis Sanitarium District	733	76	2,125,400	39
Mosquito Abatement (Four districts)	733	76	2,192,700	40

[a]Population estimated from combining population of municipalities and townships in U.S. Bureau of the Census, Census of Population: 1970. Population of that portion of South Cook County Mosquito Abatement District lying in Chicago estimated from combining population of census tracts in same source.

Problems of Definition

The term area-wide district is used to describe the Special Districts in Cook County that cover extensive areas and yet do not meet the specifications of the more commonly used term "metropolitan district."[1] According to Bollens, a metropolitan district has the following characteristics:

A metropolitan district is a special district whose territory covers a substantial part or all of a metropolitan area. . . . Whether less or more extensive territorially than the generally accepted limits of a metropolitan area, a district can accurately be called a metropolitan district only if it performs an urban function and includes the central city (or at least one central city if there are more than one) and a major part of the remainder of the territory or population of a metropolitan area.[2]

---

[1]Metropolitan district as used here refers to a type of Special District. Care must be used to avoid confusion with the term "metropolitan district" as used by the Census Bureau from 1910 to 1949 to denote a central city in addition to surrounding built-up area. The definition changed slightly from census to census, but in 1940 the term included the central city or central cities, plus adjacent minor civil divisions that had a population density of at least 150 persons per square mile. Source: Murphy, The American City, p. 15.

[2]Bollens, Special District Governments, pp. 52-53.

Pock follows somewhat the same course in defining a metropolitan district. He sets up three criteria: (1) the jurisdictional area of the district must include the core central city as well as an important part of the outlying fringe; (2) the district must perform an urban type function; and (3) the district must exhibit a high degree of administrative and fiscal autonomy.[1]

The term metropolitan district is avoided in this study because in a strict sense there are no metropolitan districts in Cook County. Both Pock and Bollens use the term for districts that do not necessarily include a major part of the metropolitan area.[2] The Census Bureau statistics do not give the population nor the area served by a Special District; therefore, unless one is very familiar with the boundary of a particular district, it is difficult to say whether it is metropolitan or not. Both Pock and Bollens mention the Metropolitan Sanitary District of Greater Chicago and Chicago Transit Authority as metropolitan districts. These districts do include a majority of the population of the Chicago Standard Metropolitan Statistical Area (S. M. S. A.),[3] but they do not include more than half of the S. M. S. A. Indeed, neither of them covers all of Cook County.

The Metropolitan Sanitary District of Greater Chicago

Nearly all of the sewage in Cook County is disposed of by plants operated by the Metropolitan Sanitary District of Greater Chicago (MSDGC). Smaller amounts of sewage are treated by other Special Districts or by a few municipalities; only 3 per cent of the population is served by privately owned septic tanks.[4]

---

[1] Pock, Independent Special Districts, pp. 12-13.

[2] Ibid.; Bollens, Special District Governments, p. 70.

[3] The Census Bureau uses the term Standard Metropolitan Statistical Area (S. M. S. A.) to define territory that includes a central city or cities in addition to territory that has metropolitan character and is integrated with the central city. The definition is rather involved, but essentially it is made up of counties, except in New England, where it is comprised of towns. For a fuller discussion, see Murphy, The American City, pp. 18-21.
  The Chicago S. M. S. A. is comprised of six counties: Cook, Lake, McHenry, Kane, Du Page, and Will. One might also include the Indiana counties of Lake and Porter which make up the Gary-Hammond-East Chicago S. M. S. A. The latter S. M. S. A. and the Chicago S. M. S. A. together form a special Census Bureau statistical area called the Chicago-Northwestern Indiana Standard Consolidated Area.

[4] Northeastern Illinois Planning Commission, Sewage Treatment, Metro-

The MSDGC has grown from its original size of 182 square miles in 1889 to its present size of 864 square miles, serving over five million people (Figure 4).

Historical Development. --The principal objective in organizing the MSDGC in 1889[1] was to make Chicago's water supply clean and safe by keeping the city's sewage out of Lake Michigan from which the city obtained its water supplies. The engineering tasks of such a project extended far beyond Chicago's borders, and a Special District was needed to have the necessary powers and territorial extent. Even then, some of the district's activities extended beyond the district boundaries.

In the decades preceding 1889 the city of Chicago was faced with serious pollution problems. Water was obtained from contaminated wells or obtained from lake supplies that were being polluted by discharges from sewers and streams emptying into the lake. Numerous attempts were made to obtain pure water by extending the intake points into the lake, but the intake points could not escape the effluent.[2]

The poor natural drainage of the Chicago area complicated many of the patchwork attempts to alleviate the pollution problem, and the demands on the political system became too great for a single municipal or township government to handle. In 1889 the Legislature approved an act creating a Special District to handle the disposal of sewage, insuring the purity of the water supply. The creation of the district was approved by the voters in an election on November 5, 1889 by the overwhelming vote of 70,958 to 242. The organization of the

---

politan Planning Guidelines, Phase One: Background Documents (Chicago: Northeastern Illinois Planning Commission, 1965), p. 8.

[1]The MSDGC is a sanitary district of the 1889 type. Laws governing sanitary districts have been passed by the Illinois Legislature for nearly a century. In identifying a sanitary district, it is common practice to designate it as belonging to a group of districts which was formed under enabling laws of a particular year. The five groups in Illinois were formed under laws passed in 1889, 1907, 1911, 1917, and 1936. Districts of the 1907 and 1911 groups do not appear in Cook County. For more details see Snider and Anderson, Local Taxing Units: The Illinois Experience, pp. 33-34.
     In this study, districts of the 1917 group are called "outlying sanitary districts" and are discussed in Chapter V as a type of Special District serving municipalities. Districts of the 1936 group are called "collector sanitary districts" and are discussed in Chapter VI as a type of Special District serving urban fringe areas.

[2]Ward Walker, The Story of the Metropolitan Sanitary District of Greater Chicago (Chicago, 1960), pp. 6-7.

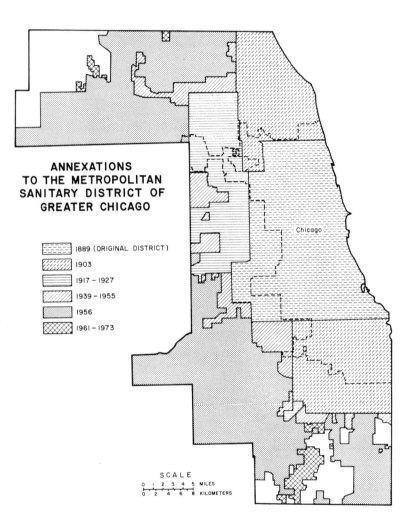

ANNEXATIONS
TO THE METROPOLITAN
SANITARY DISTRICT OF
GREATER CHICAGO

1889 (ORIGINAL DISTRICT)
1903
1917 – 1927
1939 – 1955
1956
1961 – 1973

Chicago

SCALE
0 1 2 3 4 5 MILES
0 2 4 6 8 KILOMETERS

Fig. 4

district was completed on January 18, 1890.[1]

To accomplish the task of purifying the water supply, the district decided to divert all sewage away from Lake Michigan and carry the effluent west and south to the Illinois River and ultimately to the Mississippi River. The accomplishment of this task was a considerable engineering task and required altering the natural drainage of the area.

The Chicago River, with its North and South Branches, originally flowed into Lake Michigan. In addition to these natural streams, a flat-floored valley extended to the southwest. This valley--a glacial spillway--formerly drained the glacial lake that covered the site of Chicago.[2] A canal was dug through this valley and joined to the deepened headwaters of the South Branch. By this construction, the Chicago River and the South Branch were made to flow backwards, receiving water from Lake Michigan and flowing southwest through the Sanitary and Ship Canal with which the South Branch was joined (Figure 5). Now that the pollutants were carried away from the lake, the water in the lake became safe for the water supply. All the sewers along Chicago's lake front were sealed off and their discharges made to run into the canal.[3]

Even with the reversal of the Chicago River and South Branch, the district's job was not finished. The North Branch still flowed southward, but additional water was added to the stream by construction of the North Shore Channel in 1910; this channel reaches the lake at Wilmette. On the far south side of Chicago, the Calumet River flowed into the lake bringing pollutants from industrial centers. In this case, the district utilized the same technique it had used with the South Branch. The Calumet-Sag Channel was completed in 1922 by digging a channel through another glacial spillway and joining it with the Sanitary and Ship Canal; this channel reversed the flow of the Calumet River.[4]

Sewage Disposal Facilities. --To the elaborate construction program described in the preceding section, there was added a system of sewage treat-

---

[1]Ibid., p. 7.

[2]F. M. Fryxell, The Physiography of the Region of Chicago (Chicago: University of Chicago Press, 1927), p. 18.

[3]Metropolitan Sanitary District of Greater Chicago, The Growth and Development of a Modern City (Chicago, 1962).

[4]U. S. Public Health Service, The Chicago-Cook County Health Survey (New York: Columbia University Press, 1949), p. 115.

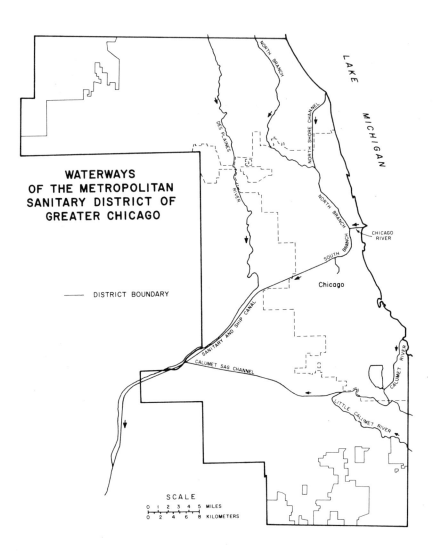

WATERWAYS
OF THE METROPOLITAN
SANITARY DISTRICT OF
GREATER CHICAGO

—— DISTRICT BOUNDARY

SCALE

0 1 2 3 4 5 MILES
0 2 4 6 8 KILOMETERS

Fig. 5

ment. The original plan called for the conveyance of all sewage to the canals and dilution with water from the lake. The diversion of water from Lake Michigan has long been a bone of contention between Illinois and the other Great Lakes States. The amount of water that was permitted to be diverted from the lake has been changed many times through the years. It began as 5,000 cubic feet per second (c.f.s.) in 1899, changed to 4,167 c.f.s. in 1901, and increased to 8,500 c.f.s. in 1925. In 1930 the U.S. Supreme Court drastically reduced the amount of diversion, reducing it in stages until by 1938 only 1,500 c.f.s. could be diverted.[1] After years of litigation, this figure was reaffirmed by a U.S. Supreme Court decision in 1967. Litigation has been complicated by the fact that water is also diverted by municipal water systems, not under the jurisdiction of the MSDGC. The 1967 decision permits domestic water system diversion of 1,700 c.f.s.--the same as the 1938 figure.[2]

The district long ago realized that with the growth of industry and population, it had to devise a system of sewage treatment for it could not expect an increasing supply of water for diversion purposes. Since the 1920's the district has erected three major sewage treatment plants in or near Chicago, supplemented by smaller treatment plants in the outlying areas. Effluent is brought to these treatment plants by a system of interceptor sewers. These sewers, owned by the MSDGC, receive liquid wastes from smaller sewer networks owned by municipalities and collector sanitary districts.

The distribution of sewage disposal facilities in Cook County illustrates the interrelationships of population density and distance. As a general rule, we may say that due to scale economies, the larger the population served by a disposal plant, the lower is the unit cost per person.[3] Also as a general rule, the greater the distance effluent must be conducted, the greater is the cost per person due to the cost of pumping stations and pipe construction.

From these two general propositions, we can see how the present sewage disposal system of the MSDGC came about. In Chicago and its contiguous densely built-up suburbs, there are three large disposal plants. Farther out from

---

[1] U.S. Public Health Service, Health Survey, pp. 117-20.

[2] A. Daniel Feldman, "The Lake Diversion Case--The End of a Cycle," Chicago Bar Record (April, 1968), pp. 270-78.

[3] For a table showing the construction cost per person of sewage treatment works according to size of plant, see Leverett S. Lyon (ed.), Governmental Problems in the Chicago Metropolitan Area (Chicago: University of Chicago Press, 1957), p. 112.

Chicago, where the population density is less, and the benefit of scale economies would not offset the increased transport costs, there are smaller disposal
plants not connected to the main plants. As the population increases, some of
the smaller plants located closer to Chicago will be eliminated and connected to
the larger plants.

Other District Activities. --The district also has dealt with problems
other than the sanitation problem. It has a hand in the shipping and navigation
problem as well. The Sanitary and Ship Canal, as one of the links in the Missis-
sippi-Great Lakes Waterway, is lined by docks and terminals.[1] The district
sewage program has necessarily included the construction of locks and dams to
regulate the water levels. The Sanitary and Ship Canal was built of suitable
dimensions for navigation, and the South Branch has been dredged and improved.
In recent years, the Calumet-Sag Channel has also been undergoing improvement for navigation.

In its history, the district had a number of activities that since have
been given up. Its police department was discontinued in 1929, and its health
department was discontinued in 1928.[2]

Legal Aspects. --The original 1889 laws have been amended many times,
and operations of the district are very restricted by legislation. The district is
governed by a board of nine trustees, of whom three trustees are elected at
large every two years for six-year terms.[3] The district may levy taxes and
issue non-referendum bonds within closely defined limits, which are frequently
modified by the Legislature.

The original powers granted the district have been increased from time
to time to permit the building and operation of sewage treatment plants, the construction of locks and dams, and, within sharply defined limits, the control of
stream pollution and flood waters. It has no authority over such activities as
water supply, construction and maintenance of local sewers, eradication of

---

[1]David M. Solzman, Waterway Industrial Sites: A Chicago Case Study,
Department of Geography Research Paper No. 107 (Chicago: University of Chicago, Department of Geography, 1966), pp. 28-47.

[2]Merriam, Lepawsky, and Sharratt, Government in the Chicago Region,
p. 56.

[3]Illinois, Annotated Statutes, ch. 42, sec. 322.

mosquitoes, or the collection and disposal of garbage and trash.[1]

Conclusions.--The MSDGC has long had the task of planning for sewage disposal in an area that is burgeoning in population and expanding in size. At the same time, technology has become increasingly sophisticated. The Legislature has permitted the district to expand its size from time to time, more or less reflecting the growth of population in outlying areas. See Figure 4 for the territorial growth of the district.

Under existing law, the district is restricted to Cook County.[2] The restriction was probably realistic in 1889, but as the Chicago area increased and population spilled over the Cook County border, the restriction has come to be less realistic. For municipalities split by the Cook County boundary, special arrangements must be made. For example, in the municipality of Hanover Park, the Cook County portion is served by the district, but the Du Page County portion is served by a municipal plant.

The MSDGC is a good example of effective planning when the political system makes it possible. Fortunately, sewage disposal is a non-controversial topic, and municipalities do not mind giving up this function. With this favorable political climate, the district can make decisions for each area without having to consider the factor of local control. Municipalities and collector sanitary districts can retain a sense of independence by local control of the sewerage network.

The Forest Preserve District of Cook County

Historical Development.--The Forest Preserve District of Cook County (hereafter called simply Forest Preserve District) is the only Special District that encompasses all of Cook County. The formation of this district was the direct outgrowth of planning in the Chicago area during the first decade of the twentieth century.[3] A commission, composed of civic leaders and prominent citizens, foresaw the need for recreational areas adjacent to Chicago as well as

---

[1] U.S. Public Health Service, Health Survey, p. 112.

[2] Laws of Illinois (1889), p. 125.

[3] William P. Hayes, "Development of the Forest Preserve District of Cook County, Illinois" (unpublished M.A. dissertation, Department of History, De Paul University, Chicago, Illinois, 1949), pp. 3-5.

such areas within the city. [1] Probably influenced by wooded areas near many
European capitals, the commission proposed an "Outer Belt of Parks and Boule-
vards" which subsequently was incorporated into the popularly known Burnham
Plan of 1909. [2]

Even though the concept of parks with forested areas had the backing of
civic leaders, the formation of the district was not easily accomplished. There
still were questions of financing and legal authorization to be answered. The
entire county evidently appeared to be the logical area for the recreational program; but due to limitations in taxation, Cook County probably would not be able
to finance the construction of the needed facilities or purchase the land. The
Illinois Constitution of 1870 stated in no uncertain terms:

> County authorities shall never assess taxes, the aggregate of which shall
> exceed 75 cents per $100 valuation, except for the payment of indebtedness
> existing at the adoption of the constitution, unless authorized by vote of the
> people of the county. [3]

In order to avoid the tax limitation and also to avoid possible legal difficulties in using the term "parks," the Illinois Legislature permitted the formation of forest preserve districts in 1905; but the legislation was deemed defec-
tive by the governor. Another act was passed in 1909, but this legislation was
declared unconstitutional by the State Supreme Court. A suitable act was finally
passed in 1913, and an election to establish a forest preserve district coterminous with Cook County was held on November 3, 1914. The proposition for the
formation of the district won easily--271, 873 votes for and 146, 895 votes against.
The district was organized on February 11, 1915. [4]

From the beginning, the objective of the district was to conserve the
rapidly disappearing woodland areas. The early officials of the district recognized the importance of acquiring forested areas before the property became too
expensive because of urban growth. In some cases, the district condemned private property for reforesting and drainage control. The acreage that the district has been permitted to acquire has been increased from time to time by the

---

[1] Special Park Commission of Chicago, Report to the City Council of Chicago on the Subject of a Metropolitan Park System, 1904 (Chicago: W. J. Hartman Company, 1905).

[2] Daniel P. Burnham and Edward H. Bennett, Plan of Chicago (Chicago:
Commercial Club of Chicago, 1909).

[3] Illinois, Constitution (1870), art. 12, sec. 8.

[4] Hayes, "Forest Preserve District," pp. 4-23, passim.

Legislature. The district is presently authorized to acquire 75, 000 acres, of which 61, 900 acres have been acquired.[1]

Although the preservation of forest areas was the district's main purpose, the district has also engaged in constructing limited recreational facilities. In addition to forest preserves, the district maintains three swimming pools, seven golf courses, and many areas for picnicking, fishing, hiking, cycling, horseback riding, camping, ice skating, and tobogganing.[2] In 1926 the Chicago Zoological Society was permitted to establish a zoological garden on the lands of the district.[3]

Legal Aspects. --The district is under the control of the Cook County commissioners acting ex officio. The commissioners meet separately to consider district affairs. Since 1926 an advisory committee, composed of interested citizens, has formulated policy for the district.[4]

According to the Census Bureau, the district is not considered as an independent government. The district is considered as a dependent district or a subordinate agency of the county government. Such a classification seems rather arbitrary, as the Illinois Supreme Court has ruled that the district is a separate legal entity. The Supreme Court has further ruled that the commissioners, when acting on forest preserve matters, are not county officers.[5]

The district is legally authorized to issue bonds and levy taxes, but these powers are under strict legislative control. The law states that districts containing over 500, 000 people shall not become indebted to the excess of three-tenths of one per cent of the assessed value of taxable property in the district. Bonds may be issued for land acquisition without voter approval, but bonds issued for other purposes must have voter approval. There are restrictions on the maximum tax rate the district may levy, as well as restrictions on the total

[1]Rutherford H. Platt, Open Land in Urban Illinois: Roles of the Citizen Advocate (De Kalb, Ill.: Northern Illinois University Press, 1971), pp. 74-77.

[2]Board of Forest Preserve Commissioners, "Welcome to Your Forest Preserve District" (Folder), n. d.

[3]Hayes, "Forest Preserve District, " p. 32.

[4]Advisory Committee to the Cook County Forest Preserve Commissioners, Revised Report (River Forest, Ill.: Forest Preserve District of Cook County, 1959), pp. 3-4.

[5]Hayes, "Forest Preserve District, " p. 32.

income of the district.[1]

Relations with Other Governments. --The territory of the district over-laps a multitude of other governments, but conflicts of authority only arise with-in the forest preserves themselves. Some of these conflicts are covered by statutes but most are resolved by mutual agreement. For example, flood con-trol measures on streams within the forest preserves involve aspects of recrea-tion, wildlife habitat, sanitation, and mosquito control. Mosquito abatement districts do not spray areas in forest preserves except by special request. The district hires rangers to maintain law enforcement within the forest preserves. However, when the rangers are acting within municipal limits, they are subject to the direction of the chief of police of the municipality.[2]

With a few exceptions, the district has not bowed to pressure to give up any of its land holdings. The district has adhered strongly to its statutory man-date which is:

> To acquire and hold lands containing natural forests, or lands connecting such forests for the purpose of protecting and preserving the flora, fauna and scenic beauties, and to restore, restock, protect and preserve the natural forests and said lands, together with their flora and fauna, as near-ly as may be in their natural state and condition for the purpose of the edu-cation, pleasure and recreation of the public.[3]

There have been recommendations for the district to develop its holdings more intensively and provide recreational facilities that would be suitable for local areas. This type of recreational development with emphasis on play-grounds, fieldhouses, swimming pools, and supervised recreation is now han-dled by park districts and municipal recreation departments. It is argued by reform groups that the fragmentation of government in the field of recreation produces waste and inefficiency.

Such development would be of doubtful legality since the district would be exceeding its legislative mandate. Even if this development would survive a court test, there are questions based on the spatial distribution of benefits and costs.

---

[1] Forest Preserve District of Cook County, "The Legislative Enabling Act to Provide for the Creation and Management of Forest Preserve Districts in Illinois," 1960. (Typewritten.)

[2] Interview with officials of the Forest Preserve District, August 19, 1970.

[3] Forest Preserve District of Cook County, Land Policy, Revised Edition (River Forest, Ill.: Forest Preserve District of Cook County, 1962), p. 11.

First of all, intensive recreation development tends to unduly benefit local areas since people do not tend to travel great distances for intensive recreation. Second, the costs of the district are borne by all of the taxpayers in Cook County. In particular, Chicago residents pay about 60 per cent of the district taxes. Third, most of the forest preserves are located in suburban areas, too far from Chicago to be used by Chicago residents if devoted to intensive recreational uses.[1]

The district makes a compromise among these conflicting conditions by emphasizing less intensively developed recreation facilities for which people are willing to travel greater distances. The facilities furnished by the district are within the distances Chicago residents are willing to travel. Suburban areas near the forest preserves can also benefit from the forested areas but not as much as they would if the district developed its holdings with intensive recreational facilities.

The Suburban Cook County Tuberculosis Sanitarium District

The Suburban Cook County Tuberculosis Sanitarium District (SCCTSD) comprises that part of Cook County, exclusive of Chicago (Figure 6). The SCCTSD, with an area of 733 square miles and a population of 2,125,400, is a relatively recent government formed only in 1949.[2] The Legislature had authorized municipalities to have tuberculosis sanitariums in 1908,[3] but authorization for a sanitarium district in combined rural and municipal territory in a county that already had a sanitarium did not come until 1937.[4]

Historical Development.--Prior to 1949 the treatment of tuberculosis in Cook County was uncoordinated; programs were administered by municipal, county, State, and private agencies. Chicago had its own municipal sanitarium, but the rest of the county was an administrative thicket of overlapping and inade-

---

[1] "Reply of Forest Preserve District of Cook County to Interim Report of the Citizens Committee on Cook County Government, March 20, 1968." (Typewritten.)

[2] Illinois City Managers' Association, Governmental Structure in the Chicago Metropolitan Area: Facts and Alternatives (Chicago: Cook County Council of the League of Women Voters, 1966), p. 22.

[3] Laws of Illinois (1909), p. 143.

[4] Illinois, Annotated Statutes, ch. 23, sec. 1701.

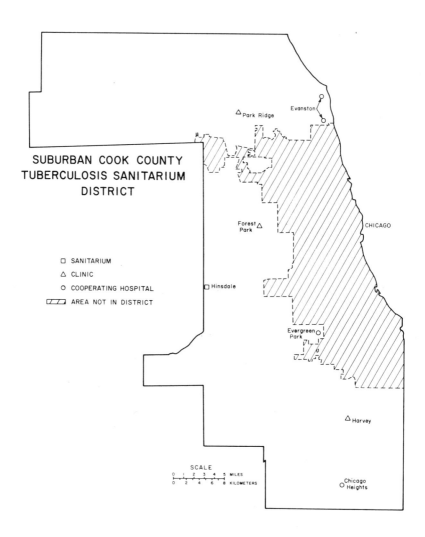

SUBURBAN COOK COUNTY
TUBERCULOSIS SANITARIUM
DISTRICT

□ SANITARIUM
△ CLINIC
○ COOPERATING HOSPITAL
▨ AREA NOT IN DISTRICT

Park Ridge

Evanston

CHICAGO

Forest
Park

Hinsdale

Evergreen
Park

Harvey

SCALE

0 1 2 3 4 5 MILES
0 2 4 6 8 KILOMETERS

Chicago
Heights

Fig. 6

quate programs.[1]

In suburban Cook County there were a few municipal health departments; but the major portion of the county was under the administration of the Cook County Department of Public Health, an agency that was too short of funds and understaffed to provide an adequate tuberculosis control program. The patients were cared for in a special section of Cook County Hospital in Chicago or at the Home for the Aged in Oak Forest. Neither facility was suitable for tuberculosis care, and patients did not want to be hospitalized in either place. The Home for the Aged possessed the stigma of a poor house, and the conditions at the Cook County Hospital were not desirable. A report notes:

> . . . The Tuberculosis Division of Cook County Hospital, having been used for many years for the hospitalization of far advanced and usually hopeless cases, is not equipped or staffed properly for even the domiciliary care given to its patients. With less than half of its potential capacity in use, actually about 170 of its beds are simply out of service. Since the public recognizes the institution for what it is, there is little likelihood that patients with a real hope of recovery could be persuaded to be hospitalized there.[2]

In addition to the county facilities, the Chicago State Tuberculosis Sanitarium was maintained by the Illinois Department of Public Health.[3] Presently, this institution is being converted to a multipurpose facility that will include treatment centers for elderly persons and heroin addicts, as well as a small treatment center for tuberculosis patients.[4]

Attempting to fill the gap in tuberculosis care provided by governmental sources, the privately run Tuberculosis Institute of Chicago and Cook County, financed largely through the sale of Christmas seals, maintained eighteen clinics and health centers in suburban areas. Even though the Institute received much of its funds from Chicago residents, it was discouraged from operating in the city by the municipal sanitarium in Chicago.[5]

The establishment of the SCCTSD was a very successful homeostatic adjustment for the political system of Cook County. The district satisfied public demands for better tuberculosis care and yet did not drastically alter the

---

[1] Robert F. Steadman, Public Health Organization in the Chicago Region (Chicago: University of Chicago Press, 1930), pp. 99-102.

[2] U.S. Public Health Service, Health Survey, pp. 518-20.

[3] Lyon, Governmental Problems, p. 184.

[4] Chicago Daily News, July 26, 1971, sec. 1, p. 2.

[5] U.S. Public Health Service, Health Survey, pp. 525-26.

existing political structure. It made it possible to coordinate the treatment of tuberculosis, equalize treatment in suburban areas, eliminate duplication of facilities, provide an adequate detection program, and provide hospital care in a specialized tuberculosis sanitarium.

The establishment of the district was politically attractive because it did not materially harm the contending political groups in Cook County. First, it permitted the suburban areas to retain their independence from Chicago. On the other hand, Chicago residents did not have to provide funds for suburban health care through their taxes to Cook County. Second, the Cook County Government did not have to increase its commitment to improved tuberculosis care, a commitment that probably would have required a tax increase or less support for some other county function.

Most municipalities in Cook County were glad to give up the care of tuberculosis patients to another level of government, owing to the expense and inconvenience of caring for these patients. In some cases, where the municipality has its own health department, the municipality cooperates with the SCCTSD in caring for patients.[1]

District Operations. --The SCCTSD maintains a hospital in pleasant surroundings in Hinsdale and also has clinics in Forest Park, Park Ridge, and Harvey (Figure 6). In addition, the district conducts weekly clinics in four suburban hospitals which are located in Evanston (two), Evergreen Park (one), and Chicago Heights (one). Four mobile vans containing X-ray units for disease detection continually travel throughout suburban Cook County.[2]

Great strides have been made in recent years in the treatment of tuberculosis. In 1950 the SCCTSD had 45 new cases per 100,000 population, but by 1968 this had fallen to 11 cases per 100,000. Similarly, the death rate for tuberculosis had dropped from 10 per 100,000 population in 1950 to 1 per 100,000 in 1968. As a basis of comparison, the state of Illinois had a death rate of 2.7 per 100,000 population in 1969.[3]

Owing to the success in combating this disease, the Legislature has permitted the district to devote some of its excess bed capacity to the treatment of

---

[1] Illinois City Managers' Association, Governmental Structure, p. 23.

[2] SCCTSD, 20th Anniversary Issue, 1949-1969 (Forest Park, Ill.: Suburban Cook County Tuberculosis Sanitarium District, n.d.), p. 17.

[3] Ibid., p. 5.

pulmonary diseases other than tuberculosis. Since taxes are collected specifically for tuberculosis treatment, the care for other diseases must be paid by the
patients themselves.[1] This modification of the original raison d'être of the district illustrates the capacity of political units to regulate their behavior and
alter their internal structure. If the Legislature were willing, it may even be
possible to modify the primary goal of tuberculosis treatment. The legal authorization for a change in goals characteristically lags far behind the actual need
or public demand.

The SCCTSD has contracts with six neighboring counties to provide care
for tuberculosis patients who live in these counties.[2] This arrangement makes
possible scale economies and a degree of specialization in tuberculosis care
that would not be possible for the individual counties. Cook County also benefits from this arrangement, since the funds received from these counties for
patient care reduces the tax burden of Cook County residents.

Legal Aspects. --The district is authorized to levy a tax up to .075 per
cent of assessed valuation in the district. Although the SCCTSD is permitted to
incur indebtedness up to one-half of one per cent of assessed valuation, it presently has no outstanding indebtedness.[3]

The district is governed by a five-member board appointed by the President of the Cook County Board with the approval of the other board members.
The district board members serve for three-year terms without compensation.
The board appoints a general superintendent who oversees the field operations,
hospital facilities, and the staff functions of the SCCTSD.[4]

Mosquito Abatement Districts

Mosquito abatement districts represent expedient solutions to problems
relating to appropriate areas, authorization of powers, and output failure on the
part of traditional governments. As urbanization proceeded in Cook County, it
began to be recognized that control measures for mosquitoes were necessary
for health reasons and to lessen the harmful effects of mosquito pests on com-

---

[1] SCCTSD, "Pulmonary Disease Wing," 1969. (Folder.)

[2] SCCTSD, "1969 Annual Report." (Folder.)

[3] Illinois, Annotated Statutes, ch. 23, sec. 1714.

[4] Illinois City Managers' Association, Governmental Structure, p. 23.

mercial and residential property values, but the governmental mechanism for performing the control measures was not clear.

County and township governments have never been concerned with mosquito control. Prior to the late 1920's, the Chicago Sanitary District (now called the MSDGC) had done much mosquito-control work through extensive drainage projects. However, in 1928 the courts ruled that the Chicago Sanitary District had exceeded its authorized powers, and it no longer could control mosquitoes.[1] Municipalities, too, were handicapped in mosquito control because many of the breeding places were outside municipal limits. Since Cook County had many swampy, poorly drained areas that afforded abundant breeding places for mosquitoes, the satisfaction of public demands on the political system became urgent. Rather than revamping the existing governmental structure, the Legislature simply authorized the establishment of mosquito abatement districts in 1927.[2]

The districts proved to be convenient mechanisms for control measures, and yet they did not threaten the existing governmental structure. The act of incorporation for these districts makes them subject to the control of municipal or other authorities. The districts do not spray in municipalities with anti-spraying ordinances, nor do they spray without permission in the holdings of the Forest Preserve District.

Four districts have been created in the county, covering about three-fourths of its area (Figure 7 and Table 9). The two oldest districts were first created in middle- to high-class suburban areas north and west of Chicago. In 1952 and 1956 with rapid urbanization in former rural areas, two more districts --much larger than the first two--were created in the southern and northwestern parts of the county. In the only district that joins part of Chicago to suburban territory, the South Cook County Mosquito Abatement District includes the southernmost part of Chicago in addition to the southern townships of the county. This part of Chicago is poorly drained and contains the bulk of the potential breeding areas in the city.[3]

The districts function effectively within their boundaries, but they are powerless to prevent the in-migration of mosquitoes from non-controlled areas. Only a very small proportion of the area of counties adjoining Cook County have

---

[1] U.S. Public Health Service, Health Survey, p. 293.

[2] Laws of Illinois (1927), p. 694.

[3] U.S. Public Health Service, Health Survey, p. 295.

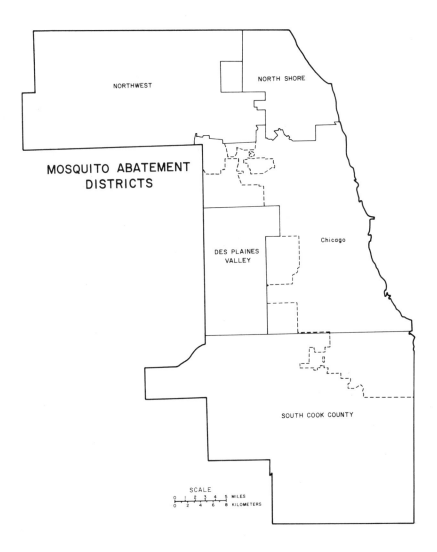

Fig. 7.--Mosquito Abatement Districts in Cook County

TABLE 9

MOSQUITO ABATEMENT DISTRICTS IN COOK COUNTY

Name	Area in Sq. Mi.	Estimated Population	Date of Incorporation
North Shore	69	320,400	1927
Des Plaines Valley	78	368,000	1928
South Cook County	349	1,017,500	1952
Northwest	237	468,800	1956

mosquito abatement districts. There are also areas in Cook County that benefit from mosquito control programs, but they are not in any district. This spill-over of benefits to areas that do not bear the cost is not unusual in Cook County.

Legal Aspects. --District affairs are managed by five trustees who are appointed by the President of the Cook County Board, with the approval of the other board members. Trustees are appointed for staggered four-year terms and serve without compensation; they must be residents of the district. Statutory provisions exist for disconnection from and dissolution of a district. The district may levy a tax up to .025 per cent of the assessed valuation of the district, but it may not issue bonds for indebtedness.[1]

Case Study: Northwest Mosquito Abatement District. --The Northwest Mosquito Abatement District was organized after the favorable outcome of an election held in April, 1956.[2] The great increase in population of the townships of northwestern Cook County had generated demands for relief from mosquito pests. The territorial convenience of the district made it possible for the townships and municipalities to cooperate in establishing the 237-square-mile district, serving residents in both incorporated and unincorporated territory.

The major efforts of the district are the elimination of breeding sources and the surveillance of known potential breeding areas. The district is continu-

[1] Illinois, Annotated Statutes, ch. 111-1/2, sec. 74-85a.

[2] Information in this case study is summarized from: Northwest Mosquito Abatement District, Cook County, Illinois, "Report on Mosquito Control Methods, 1971." (Typewritten.)

ally improving the drainage of the area by digging ditches, clearing debris from streams, and the repairing, replacing, and cleaning of drainage tiles. Inspectors keep careful check on potential breeding sources; and if mosquito larvae are present, the district sprays oil on the water surfaces of the breeding areas. As the district's territory becomes more and more urbanized, the natural breeding places disappear, but in turn many new man-made sources arise, particularly in construction sites and catch basins. A relatively small part of the district's efforts is concerned with adulticiding (killing mature insects) by the space-spray method. This technique is necessary to control adult mosquito migrations into and within the district.

Conclusions

Area-wide districts perform functions for which local townships and municipalities do not feel strong attachments. The scale of operations and degree of specialization required makes small areal units unsuitable. The large capital expenditures required in sewage disposal facilities makes local construction unattractive. Forest preserves are best handled in large tracts, and municipalities do not want huge areas of non-tax-producing land within their boundaries. Tuberculosis control requires isolation and specialized treatment facilities and would be enormously expensive if every municipality had its own facilities. Mosquito abatement takes place in widely scattered areas and is best handled by specialists not hindered by a multitude of municipal and township boundaries.

Area-wide districts are also politically attractive in that they not only avoid legal entanglements, but that they are acceptable compromises among existing political groups. As Pock says:

> The reason why metropolitan districts can be employed on a broader scale than any other remedial device is that they commend themselves to the voter and to the local politician alike; they are a moderate solution which does not interfere with the existence and the territorial integrity of other local governments, and disturbs only mildly their functional integrity. At any rate, they have come into being in many areas as a result of the failure of more extensive solutions, especially federation and county reorganization plans, to obtain popular support.[1]

There is little question that area-wide districts perform well the tasks they are legally empowered to perform. The greatest question arises as to whether they are the best mechanisms to perform their assigned tasks. In con-

---

[1]Pock, Independent Special Districts, pp. 8-9.

sideration of the legal structure, historical development, attitudes of the people, and interests of the existing political structure, the area-wide districts appear to be the most suitable mechanisms for providing certain services. In Cook County the area-wide district has proved to be acceptable to the Legislature, Cook County citizens, local officials, and the political organizations of the county. Such a consensus is a remarkable achievement. Bollens sums it up well:

> Extensive use of such alternatives as annexation by cities, city-county consolidation, and federation means the immediate displacement or substantial modification of existing governments. On the other hand, a metropolitan district with its limited functions merely adds another government to the present array and does not immediately, and in fact may never, curtail most of the services of other units. Its lack of comprehensiveness is therefore very attractive to many public officials, both elected officeholders and appointed administrators, who feel that the metropolitan district is not a potential threat to their positions, as another governmental innovation might be.[1]

---

[1] Bollens, Special District Governments, pp. 66-67.

CHAPTER V

SPECIAL DISTRICTS SERVING MUNICIPALITIES

Introduction

This chapter includes the Special Districts that, for the most part, over-lie municipal territory, although unincorporated territory is sometimes included. Some writers have used the term coterminous districts for this type of territorial coincidence, but coterminous implies a more exact areal correspondence between district and municipality than is the case in Cook County.

These districts were created as a result of the inability or unwillingness of municipal governments to provide services that were desired by inhabitants of the municipality. The reasons for this output failure of municipalities are varied, but they may be divided into three broad categories: (1) financial restrictions; (2) inappropriate areas of municipal government; and (3) the attitudes of both citizens and municipal officials. The first of these reasons is rather clear-cut and will be discussed separately; the remaining two reasons are complex and will be introduced under the discussion of individual types of districts.

In most cases, these districts provided new services for the municipality. In a few cases, existing municipal services were transferred to Special Districts.

All Special Districts described in this chapter have the power to tax, even though this power may not be exercised. Special Districts such as housing authorities, transit authorities, and port districts do not receive their funds from taxation, and they are not included in this study. Table 10 indicates the number of districts of each type, and the degree to which these districts overlie municipal territory. Each type of district will be studied individually in subsequent sections of the chapter.

TABLE 10

DISTRICTS SERVING MUNICIPALITIES IN COOK COUNTY

Type	Number	Area (Sq. Mi.)[a]	Percentage of District Area that Lies in Municipalities
Park	93	659	85
Library	14	105	69
Mass Transit	6	140	100
Outlying Sanitary	2	15	100
Public Health	2	16	88
River Conservancy	1	2	100

[a]Includes only those parts of districts which lie in Cook County.

Debt Restrictions

A very significant factor in the growth of Special Districts serving municipalities in Illinois has been the constitutional restriction on debt. This restriction severely limited the capacity of municipalities to respond to increased demands by preventing them from borrowing sufficient funds. The Illinois Constitution of 1870 specifically stated:

> No county, city, township, school district, or other municipal corporation, shall be allowed to become indebted in any manner or for any purpose, to an amount, including existing indebtedness, in the aggregate exceeding five per centum on the value of the taxable property therein, to be ascertained by the last assessment for state and county taxes, previous to the incurring of such indebtedness. Any county, city, school district, or other municipal corporation, incurring any indebtedness as aforesaid, shall before, or at the time of doing so, provide for the collection of a direct annual tax sufficient to pay the interest on such debt as it falls due, and also to pay and discharge the principal thereof within twenty years from the time of contracting the same. [1]

This restriction was not in earlier constitutions, and the apparent reason for the inclusion of such a restriction in the 1870 Constitution was a concern over the large debts which had been incurred by counties and municipalities for subscriptions to railroad stock. The proponents of the restriction felt that excessive indebtedness led to high taxes and that a restriction on debt-incurring power would serve as a means of limiting taxes. [2]

---

[1] Illinois, Constitution (1870), art. 12, sec. 12.

[2] George D. Braden and Rubin G. Cohn, The Illinois Constitution: An Anno-

It is quite evident that the framers of the constitution did not anticipate the growth of many taxing units for the same area, since they made the restriction on the units of government, not on the area involved. By the creation of several Special Districts for the same area, each with its own debt, citizens could obtain desired improvements and still not exceed the debt limit.

Many Special Districts, particularly park districts, were formed in Cook County to circumvent the debt restriction. New municipal functions were assumed by the newly created district, or, in some cases, the old municipal functions would be transferred to the district. In either case, the district would carry the debt for any new construction that might be required.

Debt restrictions will probably no longer be significant in the formation of Special Districts in Cook County. The new 1970 Constitution has given municipalities much greater freedom in acquiring debt. As the new Constitution states:

> The General Assembly may limit by law the amount and require referendum approval of debt to be incurred by home rule municipalities, payable from ad valorem property tax receipts, only in excess of the following percentages of the assessed value of its taxable property: (1) if its population is 500,000 or more, an aggregate of three percent; (2) if its population is more than 25,000 and less than 500,000, an aggregate of one percent; and (3) if its population is 25,000 or less, an aggregate of one-half percent. Indebtedness which is outstanding on the effective date of this Constitution or which is thereafter approved by referendum or assumed from another unit of local government shall not be included in the foregoing percentage amounts. [1]

As long as bond issue proposals are approved by local voters in a referendum, there is no barrier placed in the way of a home rule municipality's borrowing money to meet community demands. [2]

---

tated and Comparative Analysis (Urbana: University of Illinois, Institute of Government and Public Affairs, 1969), p. 478.

[1] Illinois, Constitution (1970), art. vii, sec. 6(k).

[2] Under the new Constitution, all cities and villages over 25,000 population automatically receive home rule status. Voters in smaller communities may obtain home rule powers through a popular referendum. The larger cities are more likely than the smaller ones to have the financial resources and the administrative expertise needed to effectively exercise home rule powers. It should be noted that Cook County has home rule status since it has a popularly elected chief executive officer (President of the Cook County Board). Source: David W. Scott, "Local Government," in Governing Illinois under the 1970 Constitution, ed. by David R. Beam (De Kalb, Ill.: Northern Illinois University, Center for Governmental Studies, 1971), pp. 28-29.

Park Districts

General Characteristics

Introduction. --Park districts are the oldest and most numerous type of
Special District in Cook County. The ninety-three park districts cover about
two-thirds of the county area and serve the bulk of its population (Figure 8).
Although municipalities were empowered to provide parks from early days, public demands for park space generally have outstripped the willingness or ability
of the municipal government to provide it. Despite the inaction of municipal
officials, local interest groups have long recognized the need for park space;
and this recognition has resulted in park district government.

The development of park districts has involved a combination of the following interrelated factors: (1) financial limitations prevented municipalities
from borrowing funds above rigid limits; (2) the conservatism of municipal officials in not providing adequate parks; (3) the desire to remove the administration of parks from the municipal political system; and (4) the areal convenience
of the district device. [1]

The constitutional limit on debt as discussed earlier surely was an
important factor in the reluctance of municipal officials to act. Nevertheless,
even when the debt limit was not reached, there has been a reluctance of municipal officials to pursue an aggressive policy of park construction. The conservatism of municipal officials is difficult to document because inaction is seldom
recorded. [2] By observing the opposite phenomenon, that is, the rise of interest
groups who form the park districts, we can infer that the municipal officials
remained inactive. A possible motive behind this inaction is suggested by Snider and Anderson who remark:

> . . . city and village government officials shy away from providing services
> that mean tax increases. Frequently, the need for referendum approval of
> a new tax levy is the death knell of a critical service. Special district devel-
> opment has been something of a soporific in that a rejected levy for city
> park purposes creates no excitement when it appears on the tax bill as a
> rate for park district purposes. [3]

---

[1] "Manual for Illinois Park Commissioners/Trustees," Illinois Parks, XII
(March-April, 1956), 33.

[2] An indication of this conservatism was obtained by scanning various
issues of the suburban newspaper, Oak Leaves (serves Oak Park and River
Forest). Of particular interest are issues for August 27, 1910; June 24, 1911;
and July 8, 1911.

[3] Snider and Anderson, The Illinois Experience, p. 13.

The desire to remove the administration of parks from the municipal system has been a tradition built up long ago in Cook County. Documentary evidence is hard to come by, but again we note that interest groups have concentrated on forming a park district rather than trying to persuade municipal officials to provide parks. In conversations with both municipal and park district officials, it seems to be an accepted situation that the administration of the two should be separate; and no sentiment was found for combining the two.

This tradition is discussed in a report concerning the proposed consolidation of the Chicago Park District and the City of Chicago.

> Conversely, it was argued that the theory of separate corporate identity for park operations, management, and maintenance was essential to the proper development of park services and facilities, which in the main seek to serve the recreational, social, and cultural needs of all the people of the community; that the history of park operation has consistently, from the time of its origin and throughout the years, been predicated upon legislative policy which foresaw the need of isolating this phase of governmental functions from the control and supervision of the municipality in which the parks were located; that park services and functions are specialized to a degree which requires the utilization of highly trained and special skills under a management independent of city government; that in terms of public necessity and welfare, the special functions of park district operation are relatively as important, if not more important, than the educational functions of a municipality and that consequently the same factors which dictated the necessity of a Board of Education free of city control are equally as valid in justification of independent status of the park district. It was pointed out that in several cities a consolidation of an independent park board with the city government has proven unsatisfactory and that in some instances where such consolidation occurred, an independent park board was subsequently re-established.[1]

The district device allows many convenient areal combinations to be set up. Municipalities may be split into more than one district, municipalities may be combined into a single district, and municipalities may be combined with unincorporated territory. The various types of combinations of municipal and park district territory will be discussed in a separate section.

Historical Development of Park Districts. --The first park districts were created in 1869 by the Legislature for Chicago and adjoining townships; these early districts are described further under a subsequent section on the Chicago Park District. It was not until after the passage of the enabling act of 1895 that smaller districts developed in Chicago and in a few suburban areas. The first district organized outside of what-is-now Chicago was the Winnetka

---

[1]"Report of the Chicago Mayor's Committee in re City of Chicago-Chicago Park District Consolidation," 1955, pp. 8-9. (Mimeographed.)

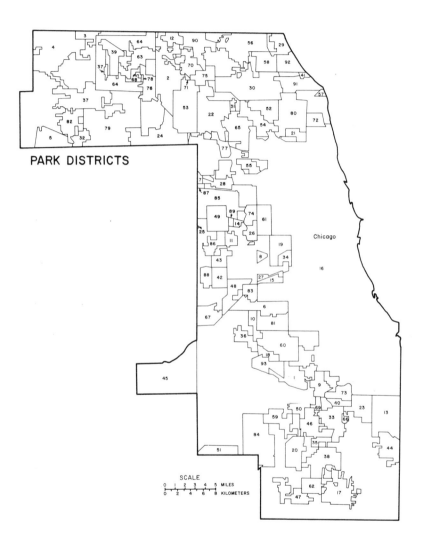

PARK DISTRICTS

Fig. 8.--Park Districts in Cook County

1. Alsip
2. Arlington Heights
3. Barrington
4. Barrington Countryside
5. Bartlett
6. Bedford Park
7. Bensenville
8. Berwyn
9. Blue Island
10. Bridgeview
11. Broadview
12. Buffalo Grove
13. Calumet Memorial
14. Central Area
15. Central Stickney
16. Chicago
17. Chicago Heights
18. Chicago Ridge
19. Clyde
20. Country Club Hills
21. Crawford
22. Des Plaines
23. Dolton
24. Elk Grove
25. Elmhurst
26. Forest Park
27. Forest View
28. Franklin Park
29. Glencoe
30. Glenview
31. Golf-Maine
32. Hanover Park
33. Harvey
34. Hawthorne
35. Hazelcrest
36. Hickory Hills
37. Hoffman Estates
38. Homewood-Flossmoor
39. Inverness
40. Ivanhoe
41. Kenilworth
42. La Grange
43. La Grange Park
44. Lanoak
45. Lemont Township
46. Markham
47. Matteson

48. McCook-Hodgkins
49. Memorial
50. Midlothian
51. Mokena Community
52. Morton Grove
53. Mount Prospect
54. Niles
55. Norridge
56. Northbrook
57. North East
58. Northfield
59. Oak Forest
60. Oak Lawn
61. Oak Park
62. Olympia Fields
63. Palatine
64. Palatine Rural
65. Park Ridge
66. Phoenix
67. Pleasantdale
68. Plum Grove Countryside
69. Posen
70. Prospect Heights
71. Prospect Meadows
72. Ridgeville
73. Riverdale
74. River Forest
75. River Trails
76. Rolling Meadows
77. Rosemont
78. Salt Creek
79. Schaumburg
80. Skokie
81. South Stickney
82. Streamwood
83. Summit
84. Tinley Park
85. Veterans
86. Westchester
87. Westdale
88. Western Springs
89. West Maywood
90. Wheeling
91. Wilmette
92. Winnetka
93. Worth-Palos

Park District in 1903.[1]

The history of these early districts is very difficult to unravel owing to a lack of records. A report of 1911 describes the situation:

> Concerning the seven smaller park districts within the City of Chicago, there is little to be said, except that their Boards are sequestered bodies of which the taxpayers know little. It was with difficulty that the Bureau was able to gather the main facts about these bodies and their business affairs. These districts were called into existence to satisfy in a crude way the needs of communities not within any other park district.[2]

In 1917 there were sixteen districts in Chicago and ten in suburban areas.[3] After 1917 suburban districts grew more rapidly than those in Chicago, totalling thirty-one by 1930; at this time, Chicago had nineteen.[4] The situation in regard to records had not improved very much by 1930 as an investigator found to his dismay. Some of his comments indicate the problem:

> With the small parks outside of cities of 30,000 population the search for data proved unexpectedly difficult and futile. A letter was sent to the secretary of each district, asking for information as called for in a blank form, and asking, in the likely event that the secretary could not conveniently supply the information required, for permission to call and examine the records. The response was disappointing. . . . The one remaining method of securing the data, through personal solicitation and by taking the information from the records, proved even more futile. In the aggregate, several days were spent in the vain effort to locate the park officials and to secure conferences with them, when located. Owing to the part-time or spare-time service of the treasurers and commissioners, they are otherwise occupied, usually away from the sites of the parks. . . . Doubtless more solicitation might have secured some additional data for some of the small park districts. But it became certain that for a considerable number of the park districts that data could not be had, no matter how much time and effort were devoted to the search.[5]

A considerable change took place in 1934 in the spatial structure of park districts as twenty-two districts in Chicago were consolidated into a single district. Suburban park districts grew slowly during the financially troubled years

---

[1]"1971 Illinois Association of Park Districts Directory," Illinois Parks and Recreation, II (September-October, 1971).

[2]Chicago Bureau of Public Efficiency, The Park Governments of Chicago: An Inquiry into Their Organization and Methods of Administration (Chicago: Chicago Bureau of Public Efficiency, 1911), p. 21.

[3]Chicago Bureau of Public Efficiency, Unification of Local Governments in Chicago (Chicago: Chicago Bureau of Public Efficiency, 1917), pp. 21-23.

[4]Jens Peter Jensen, "Financial Statistics of Governments in Cook County" (unpublished report, University of Chicago, 1931), pp. 4-31.

[5]Ibid., pp. 226-27.

of the 1930's and reached the number of thirty-six in 1939. It was the great out-
migration of population from Chicago and the older suburbs after World War II
that stimulated the greatest growth of suburban park districts. Table 11 shows
a summary of the growth of park districts for the past century. Figure 9 shows
the areal growth of park districts. We note that the growth of park districts
has followed the outward growth of population from Chicago.

TABLE 11

PARK DISTRICTS IN COOK COUNTY, 1869-1973

Year	Number[a]	Area[a]	
		Sq. Mi.	Per Cent of County Area
1869	3	136	14
1889	3	146	15
1917	26	211	22
1930	50	--[b]	--[b]
1939	37	343	36
1962	59	--[b]	--[b]
1973	93	659	69

[a]See Appendix II for sources.

[b]Information not available.

Looking at the development of park districts as a historical process, we
observe a basic pattern that has been repeated time and time again. As an area
was converted from rural land use to urban land use, the very rapid growth in
population exerted considerable pressure on the existing park facilities. If the
area was incorporated, there was usually a long delay in the provision of park
facilities by the municipal government. If the area was unincorporated, township government was not geared psychologically or financially to construct park
facilities.

As demands mounted on the traditional political system, local interest
groups arose which desired recreation facilities immediately. Leaders of these
groups recognized that recreation land had to be purchased quickly before it
became built-up or too expensive. Rather than trying to change the attitude of
municipal officials, it was far simpler to petition the Circuit Court for an elec-

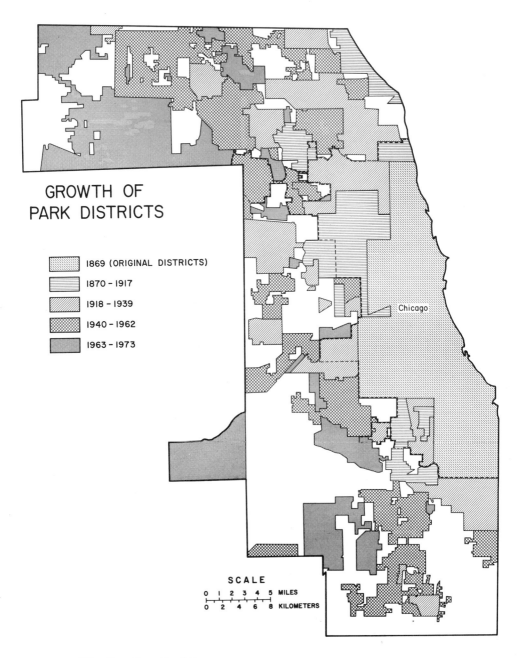

GROWTH OF
PARK DISTRICTS

1869 (ORIGINAL DISTRICTS)

1870 – 1917

1918 – 1939

1940 – 1962

1963 – 1973

Chicago

SCALE

0 1 2 3 4 5 MILES

0 2 4 6 8 KILOMETERS

Fig. 9.--Growth of Park Districts in Cook County, 1869-1973

tion to establish a park district, an endeavor which has received considerable voter approval.

The abundance of park districts in Cook County reflects a non-coordinated approach to residential development. There seems to be no orderly mechanism to set aside land for park purposes by municipal or township government officials. In only a few cases, has the municipal government (such as South Holland) required land to be set aside for parks by developers. Park Forest is exceptional in that it was a planned community from the beginning, and allowances were made for recreation land to be set aside. Even in those areas, in which a municipality is incorporated after a park district has already been established, there is little sentiment to change the existing park district government. Citizens evidently prefer the individual attention the park district provides; they feel it is more responsive to their needs than a municipal bureau of recreation.

Legislative History. --In 1893 the first enabling legislation for park districts was passed by the Legislature in a bill entitled "an act to provide for the creation of pleasure driveway and park districts."[1] In 1895 the Legislature passed a second general enabling act entitled "An act to provide for the organization of park districts and the transfer of submerged lands to those bordering on navigable bodies of water." Actually, it was not necessary under the act for each park district to border on navigable water bodies, but park districts formed under this act were known as Submerged Land Park Districts. Except for the Chicago Park District and the Oak Park Park District, all park districts in Cook County formed prior to 1947 were formed under this act.

In 1911 the Legislature passed a third general act for the establishment of parks and parkways by towns and townships. The park district in Oak Park is the only one in Cook County to have been formed under this act.[2] The Park District Code, enacted in 1947 and amended in 1951, made the laws governing park districts relatively uniform. Under this code, the only type of park district that can be formed in the future is a General Park District. The code applies only to park districts of less than 500,000 inhabitants, and it does not contain the laws concerning the maintenance of parks by cities and villages. The code does not affect the Chicago Park District (more than 500,000 population), which has a separate body of laws.

---

[1] The section on legislative history is adapted from: Illinois, Annotated Statutes, ch. 105, "History of the Park District Code," pp. xiv-xix.

[2] Oak Leaves, July 8, 1911, p. 3.

The park district is managed by a park commission consisting of five unsalaried members who are elected for six-year terms. Tax levies are determined by the commissioners within a stipulated maximum, but this maximum can be increased by referendum.

Spatial Arrangements of Park Districts and Municipalities. --The area flexibility of the district device leads to considerable variation in the arrangement of park districts and municipalities. In most cases, the park district and municipality grew up independently, with little regard for the eventual problems of coordination that would result. For convenience in discussion, the following types of arrangement will be discussed: (1) municipality and park district are coterminous or nearly so; (2) the park district occupies only a small part of the municipality; (3) the park district contains surrounding unincorporated territory in addition to the municipality; (4) the park district contains several municipalities; (5) the park district lies principally in unincorporated territory.

The most common type of arrangement is the situation in which the municipality and park district have approximately the same area. At one time, they may have had exactly the same area; but since an expansion of one is not necessarily followed by a corresponding change in the other, they often are slightly different. Only in Chicago does the park district boundary automatically change with the municipal boundary. In these cases of great areal coincidence, the park district simply handles most of the recreation facilities of the municipality. Even though legally separate, the park district and municipality operate in a cooperative fashion. Through time a series of understandings has arisen, and both municipality and park district enjoy popular support.

Before World War II many small park districts were created entirely within municipalities, providing services to municipal neighborhoods. Prior to 1934, Chicago had twenty-two park districts within its borders. Small districts of this type still remain in a few of the older suburbs, such as Evanston, Berwyn, and Maywood. These neighborhood districts are supplementary to recreational facilities maintained by the municipal government. Although their small size greatly limits the tax resources available to these small districts, they offer types of programs and permit a form of intimacy not found in larger park districts or municipal facilities.

The third type of park district-municipal arrangement is one in which the district includes unincorporated territory in addition to the municipality. This situation benefits citizens who live close to the municipality and use park district facilities, but do not wish to become part of the municipality. The service area of the district does not necessarily correspond with the municipality; yet the district device permits residents of unincorporated areas to enjoy urban-type services and pay taxes for these services.

Districts serving more than one municipality are the fourth type of park district-municipality arrangement. The districts permit municipal cooperation in the provision of park facilities; yet the municipalities can maintain their independence in other matters. These districts are large enough to have an adequate tax base and offer a variety of recreation programs; yet they are small enough to be responsive to citizen demands.

The last type of areal combination concerns park districts that were organized primarily in unincorporated territory. These districts are more properly called "fringe districts," which are discussed at length in the next chapter. They are suitable for those citizens who want recreation facilities but do not want municipal government. Actually, the unincorporated stage is usually temporary, and there is no park district in Cook County that does not contain some municipal territory within its boundaries.

Case Study: Mount Prospect Park District

Introduction. --The formation of the Mount Prospect Park District is typical of the recent growth of park districts in Cook County. Sudden stress caused by greatly increased demands on the existing political structure resulted in the creation of a new governmental form to minimize the stress. The existing municipal park system was unable to accommodate the greatly increased demands of a burgeoning population. The formation of the park district satisfied the demands for recreational facilities and at the same time did not greatly alter the existing political forms.

Historical Development. --The village of Mount Prospect is a high-income dormitory suburb, lying northwest of Chicago. The village was incorporated in 1917, but it grew very slowly before World War II.[1] In the decade after the war, the village grew rapidly, increasing in population from 4,000 in 1950 to 12,000 by 1954. The village had a sixteen-acre park and six smaller ones, ranging from one to five acres. This acreage, however, was completely

---

[1]The historical account is taken from: "Mount Prospect District Approves Master Plan," Illinois Parks, XII (March-April, 1956), 29-32.

inadequate for the demands of the fast increasing population. The neighboring park districts were also pushed to capacity and could not accommodate the overflow of Mount Prospect residents. The village did not have the bonding power to incorporate a swimming pool and parks into its program. The formation of a park district with separate bonding power seemed to be the answer.

After a postcard survey had determined that support for the district would be forthcoming, the district was approved in an election on June 4, 1955. A bond issue for $450,000 was passed in late 1955 and a 4 per cent recreation tax on taxable property in the district was approved. The district included considerable territory in addition to the village of Mount Prospect. The added territory allowed for new residential development, but more important than the room for expansion was the fact that areas in the southern part of the district provided potential industrial sites. The southern part of the district is near the O'Hare Airfield of Chicago, and good transportation is provided by expressways. In Cook County the inclusion of tax-rich industrial property is desirable because it permits the district to benefit from added revenues and yet requires little additional expenditures to provide services. The foresight of the decision-makers was accurate, and the southern part of the district is now in the industrial area of Elk Grove Village (Figure 10).

Present Conditions. --The district has a population of about 50,000 in an area of eleven square miles. The district owns 272 acres of park land and leases an additional 61 acres. The district also has a fieldhouse, three swimming pools, and a golf course (Figure 10).

The district now includes most of the village of Mount Prospect and parts of the municipalities of Arlington Heights, Des Plaines, and Elk Grove Village. The association of district residents from these municipalities has created a loyalty to the district, and there is no sentiment to make the district boundary coincide with municipal boundaries.[1]

Case Study: Chicago Park District

Introduction. --The Chicago Park District is the giant of Cook County park districts. Not only is the district unique because of its great size but also by the scope of the programs it offers. The district has had a long and complex history, involving the interaction of scores of actors and political groups.

---

[1]Interview with Director of Mount Prospect Park District, August 9, 1971.

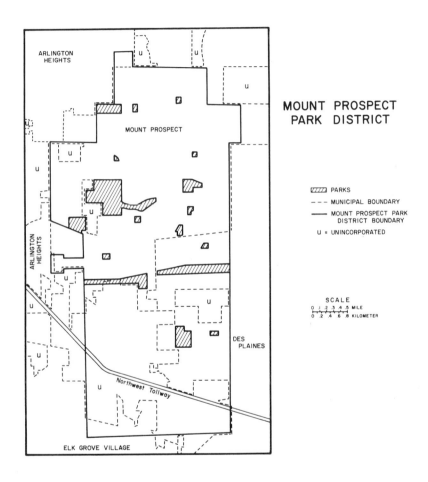

MOUNT PROSPECT PARK DISTRICT

Legend:

▨ PARKS

- - - MUNICIPAL BOUNDARY

——— MOUNT PROSPECT PARK DISTRICT BOUNDARY

U = UNINCORPORATED

SCALE
0 | 2 3 4 .5 MILE
0 2 4 6 .8 KILOMETER

Fig. 10

The Chicago Park District is coterminous with the city of Chicago; and, unlike other park districts, the boundary of the district changes with that of the city. The political control of the district is highly centralized. Unlike other park districts, where the commissioners are elected, the mayor of Chicago appoints the five-member Board of Commissioners who serve without pay for staggered five-year terms.[1] Within Chicago, the district has control of most, but not all, of the recreational facilities of the city. Although the city of Chicago has transferred most of its recreational facilities to the district, some recreational facilities are still under the Chicago Board of Education and the Forest Preserve District.

The Chicago Park District was achieved through the consolidation of three large and nineteen small park districts in 1934. This consolidation, discussed at length in a later section, represents one of the few cases in Cook County where consolidation of Special Districts has taken place.

Historical Development. --In 1869 the Legislature created three park districts for Chicago and adjoining townships.[2] This division marked the beginning of separate governments for parks, and this separation was strengthened by the passage of general enabling legislation for park districts in 1893 and 1895. Evidently, the Chicago municipal government did not provide sufficient parks, and interest groups succeeded in having the Legislature pass bills for creating separate park governments.

All three districts were created by private acts of the Legislature; all three were passed in a single month; and all three were different. It appears there was a desire to assess the costs of developing the parks in those areas that would benefit from their existence.[3] These areas did not coincide with Chicago's boundaries; so apparently the Legislature created districts that would nearly match benefits and costs. There was considerable jealousy among sections of Chicago in the late 1860's, and there was considerable fear among park proponents that the public would reject bills that taxed one area and benefited another.[4]

---

[1] League of Women Voters, Key to Local Government, p. 51.

[2] Except where otherwise noted, this section on historical development is taken from Merriam, Lepawsky, and Sharratt, Government in the Chicago Region, pp. 44-52.

[3] "The South Side Park Bill," Chicago Tribune, January 25, 1869, p. 2.

[4] Everett Chamberlin, Chicago and Its Suburbs (Chicago: T. A. Hungerford & Company, 1874), p. 313.

Figure 11 shows the geographical setting of the three park districts. By far the largest district was the South Park District for the southern part of Chicago as well as adjacent Lake and Hyde Park Townships. Lincoln Park District was designed to service the northern part of Chicago and adjacent Lake View Township. The West Chicago Park District served the western part of the city.[1] In 1889 the townships of Lake, Hyde Park, and Lake View were annexed to Chicago, and all three park districts were completely within the city. Part of the raison d'être for creating the park districts was to join areas inside and outside Chicago; yet twenty years after creation of the park districts, all three were completely within the city. During the first twenty years of their existence, the districts had attracted substantial support, and the tradition of separate park district government had been established.

The enabling legislation for park districts in 1895 enabled small park districts to be formed by local initiative in those sections of Chicago not covered by the three original park districts. In 1917 there were thirteen small districts in addition to the three large ones. In addition to parks maintained by the park districts, the city of Chicago, acting through its Special Parks Commission, also maintained many small parks and playgrounds within the areas of the park districts.[2]

Among the three large park districts in Chicago, great inequalities in wealth existed. The South Park District was in the best strategical position, since it covered the tax-rich downtown area as well as the southern industrial areas of the city. In 1931 South Park District contained 48 per cent of the assessed valuation of Chicago, but it contained only 36 per cent of the city's population. The West Chicago Park District, on the other hand, included only 19 per cent of the city's wealth, but served 29 per cent of the city's population. The Lincoln Park District was in an intermediate position, containing 19 per cent of the assessed valuation and serving about 14 per cent of the population. The West Chicago Park District embraced some of the neediest areas of the city; its hous-

---

[1] West Chicago Park District was coincident with West Town (township). As Chicago annexed territory from the Town of Cicero, this territory was transferred to West Town and automatically became part of West Chicago Park District. In some sources the district is named simply West Park District.

[2] Chicago Bureau of Public Efficiency, The Nineteen Local Governments in Chicago: A Multiplicity of Overlapping Taxing Bodies with Many Elective Officials (Chicago: Chicago Bureau of Public Efficiency, 1915), p. 28. A similar situation exists today in Evanston. The city of Evanston maintains parks inside the small park districts within its borders. Source: City of Evanston, Planning and Conservation Department, "Parks and Park Districts," n.d. (Map.)

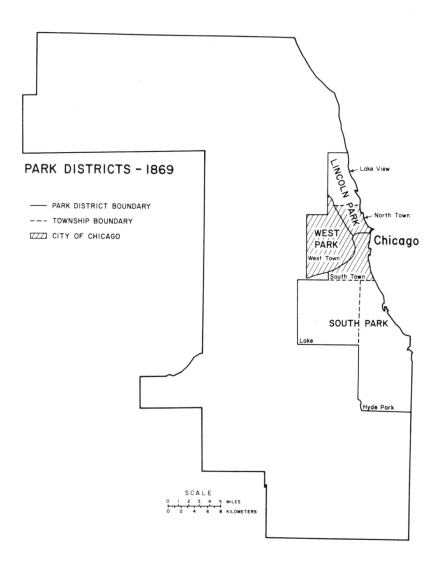

PARK DISTRICTS - 1869

—— PARK DISTRICT BOUNDARY
- - - TOWNSHIP BOUNDARY
▨ CITY OF CHICAGO

LINCOLN PARK

← Lake View

← North Town

WEST PARK

Chicago

West Town

South Town

SOUTH PARK

Lake

Hyde Park

SCALE

0 1 2 3 4 5 MILES
0 2 4 6 8 KILOMETERS

Fig. 11.--Three Original Park Districts in Chicago, 1869

ing conditions made its park needs the greatest, yet it could afford and offer the least. Thus the presence of three large park districts in the city resulted in gross inequities among the distribution of park facilities and failed to strike a balance between resources and requirements.

Dissimilarity in resources among the three large park districts was accompanied by variations in their political relations with other governments. Although the three park systems were created at the same time, there seems to have been no attempt to set up a uniform plan of political responsibility to the citizens of Chicago. The West Chicago Park Commissioners were appointed by the governor of Illinois. On the other hand, the Lincoln Park Commissioners were named by the Legislature in the original act, and their successors were to be appointed by the Circuit Court of Cook County. In 1871, however, the act was changed to vest the appointment power with the governor. The original South Park Commissioners were appointed by the governor, but their successors were to be named by the Circuit Court.

Whereas the South Park District was administered by a board of five commissioners, juggled into office through political factions represented on the Circuit Court, the West Chicago Park District and Lincoln Park Districts were under the administrative control of a board of seven commissioners designated by the governor. The commissioner jobs were political plums. The governor held, through this appointing power, the key to a tremendous patronage-dispensing organization in the heart of Chicago. Through his position, he had been able to command the support of local representatives and senators. A governor merely had to threaten to take, give, or redistribute the fruits of this potentially powerful patronage, and certain factions and individuals became his political allies.

In addition to these three major park districts, there were nineteen other small park districts in Chicago in 1933.[1] Since these districts were created under the enabling legislation of 1895, they had locally elected commissioners, and thus were able to escape the patronage systems of the governor and Cook County Circuit Court. In 1931 these smaller districts, covering about one-fourth of Chicago's area, contained 18 per cent of its population and 14 per cent of its assessed property. The districts varied widely in size and population, ranging in

---

[1]Unfortunately, no map has been found that shows all the park districts in Chicago in 1933. The closest approximation to the actual pattern appears in the previously cited Chicago Bureau of Public Efficiency, Nineteen Local Governments in Chicago, p. 27. This 1915 source has a map showing the three large districts and nine of the smaller ones.

size from one-half square mile to ten square miles. These small districts
were poor in resources in proportion to their population and area, and some
were too small to provide even the most meager park facilities.

Creation of the Chicago Park District. --The consolidation of the three
large and nineteen small park districts in Chicago in 1934 into a single district
is the only substantial consolidation of Special Districts that has taken place in
Cook County. In order for this consolidation to have taken place, there had to
be great stress placed on the political system, and the system had to be sufficiently organized for amelioration of the stress to take place. Both conditions
were present in the depression days of the early 1930's. In the first place, it
was a time of great financial distress. Tax delinquencies were widespread,
many park districts had defaulted on their bonds, and the appeal of lower taxes
was great. Second, the Democrats had acquired political control in Chicago in
1931, and the city was ruled by an efficient patronage-dispensing machine.[1]

The political leaders of Chicago evidently perceived that the consolidation could not be brought about by having the elections held separately in each
small park district, where local loyalties would frustrate the creation of support for an all-Chicago district. On the other hand, the three large park districts never had elected their own officials; consequently, the loyalties in these
districts could be more easily marshalled into support for an all-Chicago district. Hence, if the election could be arranged in such a manner that the vote in
the large districts could be counted in with the vote in the small districts, the
support for the all-Chicago park district would probably win out. Special legislation was passed by the Legislature in 1933 to permit an election to be held in
all of Chicago to form a single park district.

In addition to the change in loyalties that would be required, the proposed consolidation would involve a realignment of power and patronage in Chicago. Not only would all the locally elected park commissioners in the small
districts have to be supplanted, but the governor and Circuit Court would lose
their patronage power and transfer it to the mayor of Chicago.

The period before the election was a period of great controversy. The
proposed park consolidation was not a clear-cut case of efficiency versus inefficiency; sound arguments could be advanced both for and against. On the one
hand, unified control would allegedly lead to a more efficient use of resources,

---

[1]Victor Jones, "Local Government in Metropolitan Chicago, " American
Political Science Review, XXX (October, 1936), 935.

duplication of facilities would be eliminated, and total costs would be reduced.[1] On the other hand, if individual parks were kept, the local leaders would be more familiar with the needs of their constituents, the smaller districts would remain free from political manipulation, and taxes would be lower because greater control could be exercised over officials by beneficiaries of the services.[2] Strangely enough, arguments about inequalities of resources and needs were not mentioned.

The election was held on April 10, 1934, and the proposition for consolidation carried by a vote of 507,955 to 174,631. A majority for consolidation was obtained in the three large districts and in some of the small districts. Adverse votes were cast in most of the small districts. In August, 1934 the Illinois Supreme Court ruled that the entire procedure was legal, maintaining that the Legislature had the power to abolish or consolidate any municipal corporation at its discretion.[3]

The fledgling Chicago Park District was faced with a host of problems for many of which there existed no precedent. The South Park District was in the best financial position, and its resources helped bail out the rest of the city. Most districts were defaulting on the interest and principal of their bonded indebtedness. The new district assumed all the indebtedness of the former individual districts, and a beginning was made to redistribute the available resources.[4] In later years, the indebtedness was cleared up; moreover, great disparities among individual parks were lessened. As noted in a report:

> The advantages of park consolidation are not reflected solely in dollars and cents. The first year of unified operations has already shown increased usefulness of park facilities and greater benefits for all our citizens. Many recreational and service features have been established at numerous parks where they were not previously available.[5]

The Chicago Park District today provides an impressive program of recreational facilities. The district has nearly 7,000 acres of parks, thirty

---

[1]"Prepare for Drive to Win Votes for Merger of Parks," Chicago Tribune, March 3, 1934, p. 7.

[2]James B. Kenny, "Development in Illinois Park Districts: With Special Reference to the Small Park Districts in Cook County," Illinois Municipal Review, XIII (January, 1934), 11-12.

[3]Jones, "Local Government in Metropolitan Chicago," p. 937.

[4]Ibid.

[5]Chicago Park District, First Annual Report: May 1, 1934 to December 31, 1935 (Chicago, 1936), pp. 10-11.

bathing beaches, and many park fieldhouses that provide a variety of supervised recreation programs. Many of its specialized facilities such as museums, a zoo, a planetarium, and an aquarium are regional in character; that is, they are utilized by many people in addition to Chicago residents. Whether local programs have been short-changed in order to provide these regional facilities is a question worthy of investigation.

Evaluation of the Spatial Structure
of Park Districts

Resources and needs for recreation are spread unevenly throughout Cook County outside Chicago, a situation similar to the one that existed in Chicago before 1934. Unlike the area-wide districts discussed in the previous chapter, in which differences in wealth among micro-areas were insignificant, the park districts present great differences in available resources because of their limited size.

Since park districts receive most of their income from taxation, the assessed valuation of taxable property is a good indicator of resources available to the park district. Table 12 ranks the districts according to assessed valuation per capita; such a figure eliminates variations due to size and population of districts. The table shows that a few districts have enormous resources compared to the bulk of the districts. These higher ranked districts either contain extensive industrial property or are very high-class residential areas. The average per capita assessed valuation for all citizens in park districts is $3,427.

Since there is no common measure for quality of services, one cannot rank the districts on this factor. Even the available resources are not a true indicator because the tax rate is variable. Visits to park districts have shown a great variation in facilities and staff. At the upper end of the scale, districts offer swimming pools, golf courses, fieldhouses, and a large staff of recreation specialists offering many programs. At the lower end of the scale, districts may have almost no full-time employees, a minimum of facilities, and offer almost no supervised activity. Districts in wealthy areas have many activities for local residents and charge fees that would be beyond the means of residents of poorer areas.

In effect, the present park district organization permits great inequalities to exist in recreational facilities. This discrepancy reflects the minimal commitment of Cook County citizens to egalitarian goals. Even in the poorer districts, which would benefit most from enlarged areas, little sentiment for consolidation was noted. Except for regional facilities in Chicago, park facilities are utilized largely by people who live close by. Therefore, the small size of park districts outside Chicago permits a high quality of services to be maintained in wealthy areas and a much lower quality in poorer areas. The spatial structure also permits those districts with large tracts of industrial property within their borders to benefit from large tax resources, without having to render any additional services.

Another problem in matching resources and recreation needs arises from the regional facilities of the Chicago Park District. This district, which is slightly below average in assessed valuation per capita, maintains facilities that are used by the citizens of all of Cook County and beyond; yet these facilities are paid for by Chicago taxpayers. Clearly, if residents of suburban park districts could not use the Chicago Park District regional facilities, the suburban park districts would have to devote more of their resources to specialized facilities.

Public Health Districts

Only two public health districts, Berwyn and Stickney, have been created in Cook County (Figure 12). Both are located in suburban areas close to Chicago's western boundary. Their 1970 area totaled 16.1 square miles, and their population was about 94,000.

The formation of public health districts is authorized in a 1917 act of the Legislature. These districts are coextensive with townships, and the district's governing officials are the regular township officers, acting in an ex officio capacity. They exercise independent taxing and governing powers and may issue tax anticipation warrants. The district may levy a tax not to exceed .066 per cent of the full cash value of taxable property in the district.[1] The Census Bureau considers these districts as dependent districts, and they are not listed in tabulations of the Census Bureau.

Case Study: Berwyn Public Health District.--The Berwyn Public Health District, organized in 1920, is coextensive with the township and city of Berwyn (52,202 population in 1970). In practice, the district operates as a municipal board of health and is sometimes called the Berwyn Health Department. The district provides communicable disease control, maternal and child health care,

---

[1]Illinois, Annotated Statutes, ch. 111-1/2, sec. 1-15.

TABLE 12

RESOURCES OF COOK COUNTY PARK DISTRICTS

District	Assessed[a] Valuation (000's) $	Population[a]	Assessed Valuation Per Capita $	Rank
Forest View	56,611	1,000	56,611	1
Bedford Park	131,640	3,000	43,880	2
Hodgkins-McCook	81,575	3,300	24,720	3
Rosemont	43,862	3,500	12,532	4
Central Stickney	28,149	2,800	10,053	5
Schaumburg	63,657	6,500	9,793	6
Kenilworth	25,823	3,000	8,608	7
Broadview	76,692	9,500	8,073	8
Glencoe	73,017	10,500	6,954	9
Elk Grove	136,115	20,000	6,806	10
Franklin Park	120,049	18,000	6,669	11
Winnetka	104,350	16,000	6,522	12
Barrington Countryside	17,274	2,712[b]	6,369	13
Hawthorne	43,034	7,000[c]	6,148	14
Olympia Fields	18,176	3,000	6,059	15
River Forest	75,723[d]	12,500	6,058	16
Barrington	46,177	7,900	5,845	17
Northfield	28,925	5,000	5,785	18
Northbrook	106,943	19,160	5,582	19
Lemont	26,420	5,000	5,284	20
Skokie	375,437	72,000	5,214	21
Bensenville	71,901	14,000	5,136	22
Wilmette	158,851	31,685	5,013	23
Morton Grove	122,432	27,000	4,535	24
Niles	158,055	35,000	4,516	25
Bartlett	16,220	3,600	4,506	26
La Grange	76,698	17,600	4,358	27
Veterans	290,992	68,000[e]	4,279	28
Clyde	255,384	60,000	4,256	29
Glenview	180,746	42,470	4,256	30
Elmhurst	197,244	48,000	4,109	31
Homewood-Flossmoor	97,388	25,000	3,896	32
Alsip	36,437	9,800	3,718	33
Western Springs	48,285	13,000	3,714	34
Norridge	66,440	18,000	3,691	35
Inverness	6,181	1,700	3,636	36
Wheeling	50,673	14,000	3,620	37
River Trails	46,720	13,000	3,594	38
Oak Park	225,580	63,000	3,581	39
Chicago Heights	150,048	42,000	3,573	40
Des Plaines	201,352	57,000	3,532	41
La Grange Park	56,326	16,200	3,477	42
Summit	36,265	10,500	3,454	43
Mount Prospect	189,365	55,000	3,443	44
Berwyn	32,495	9,500	3,421	45
Riverdale	51,901	15,706[e]	3,305	46

TABLE 12--Continued

District	Assessed[a] Valuation (000's) $	Population	Assessed Valuation Per Capita $	Rank
Memorial	160,307	50,000	3,206	47
Dolton	70,262	22,000	3,194	48
Chicago	11,442,669	3,587,000	3,190	49
Westdale	4,781	1,500	3,187	50
Arlington Heights	190,052	60,000	3,168	51
Harvey	106,650	35,000	3,047	52
Westchester	66,159	22,000	3,007	53
Rolling Meadows	53,476	18,000	2,971	54
Prospect Heights	20,753	7,000	2,965	55
Blue Island	65,090	22,000	2,959	56
Forest Park	45,580	15,750	2,894	57
Calumet Memorial	97,973	34,000	2,882	58
Oak Lawn	164,500	58,800	2,798	59
Worth-Palos	29,514	11,000	2,683	60
Bridgeview	30,509	11,500	2,653	61
West Maywood	11,359	4,500	2,524	62
Phoenix	9,051	3,596[b]	2,517	63
Matteson Memorial	13,297	5,300	2,509	64
Chicago Ridge	22,480	9,000	2,498	65
Palatine	65,089	26,100	2,494	66
Markham	35,260	14,500	2,432	67
Hoffman Estates	46,617	20,000	2,331	68
Tinley Park	27,914	12,000	2,326	69
Country Club Hills	13,277	5,838	2,274	70
Hickory Hills	31,626	14,000	2,259	71
Golf-Maine	24,663	11,000[e]	2,224	72
Oak Forest	30,802	14,300	2,154	73
Posen	11,281	5,300	2,128	74
South Stickney	60,178	30,000	2,006	75
Midlothian	28,177	15,000	1,878	76
Streamwood	28,590	16,000	1,787	77
Hanover	15,430	10,300	1,498	78
Total	17,526,994	5,114,417		

For notes and sources, see Appendix II.

school health care, milk and food sanitation control, and rodent control.[1]

The initial raison d'être of this governmental form is not obvious from present conditions. A 1933 study suggests that at the time when townships in

[1]Lyon, Governmental Problems, p. 169.

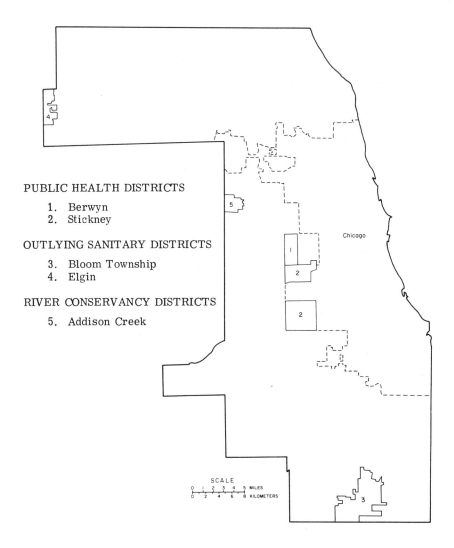

PUBLIC HEALTH DISTRICTS

   1. Berwyn
   2. Stickney

OUTLYING SANITARY DISTRICTS

   3. Bloom Township
   4. Elgin

RIVER CONSERVANCY DISTRICTS

   5. Addison Creek

Fig. 12.--Public Health, Outlying Sanitary, and River Conservancy Districts in Cook County.

Cook County assessed their own property, Berwyn Township wanted to keep its relatively low assessment ratio and thereby keep down its local tax rate. By keeping the ratio of assessment to actual value at a low figure, Berwyn Township was able to minimize its proportional share of Cook County taxes.[1] Real property in Cook County is now assessed by county officials, and the township assessors are deputies of the county assessor. The original raison d'être of the district has long passed, but the governmental form still persists. To the residents of Berwyn, it evidently makes little difference whether the health officials are township or municipal officials.

Case Study: Stickney Public Health District. --Immediately south of the Berwyn Public Health District lies the Stickney Public Health District, organized in 1946. This district comes close to the original purpose of the legislative act, which was to provide health services for several municipalities that might lie within a township.[2] Stickney Township contains the municipalities of Forest View, Stickney, Burbank, and parts of Bridgeview and Bedford Park. The population of 41,752 benefits from large nearby industrial areas that provide millions of dollars of taxable property, yet demand a minimum of health services. The municipalities do not have to establish health departments and can maintain low tax rates.

The district provides a well-rounded program of environmental sanitation, maternal and child care, including school programs, communicable disease control, and partial public health laboratory services.[3]

Outlying Sanitary Districts

The outlying sanitary districts are defined as those districts organized under the sanitary district law of 1917. These districts have disposal plants which treat sewage collected by municipal sewerage systems. The two districts of this type in Cook County were organized before the MSDGC was organized in these areas. Unlike the MSDGC, there is no legal restriction about crossing the Cook County boundary for sanitary districts of the 1917 type. The Bloom Township Sanitary District, organized in 1928, lies in Cook and Will Counties; the

---

[1] Merriam, Lepawsky, and Sharratt, Government in the Chicago Region, p. 61.

[2] Steadman, Public Health Organization, p. 30.

[3] Lyon, Governmental Problems, p. 169.

Elgin Sanitary District, organized in 1922, lies in Cook and Kane Counties (Figure 12).[1]

Both of these sanitary districts were created to avoid the constitutional limit on debt. Both Chicago Heights (in Bloom Township) and Elgin had to give up their disposal plants to sanitary districts because they could not legally borrow money for needed plant improvements. The law permits these districts to include more than one municipality and to include unincorporated territory within one-and-one-half miles of municipal boundaries.[2]

Legal Aspects. --These districts are managed by three trustees who are appointed by the chief executive officer of the county in which the greater part of the district's territory lies. These districts may borrow funds and levy taxes up to .083 per cent of the assessed valuation of property in the district.[3]

Case Study: Bloom Township Sanitary District. --The Bloom Township Sanitary District encompasses an area of about 14 square miles in southern Cook County and northern Will County. The district treats the raw sewage collected by the municipal sewerage systems of Chicago Heights, Park Forest, and South Chicago Heights. In 1970 the total population of these municipalities was 76,400.[4]

The district was formed as a result of output failure on the part of municipal government. The sewage treatment plant of Chicago Heights could not be improved because the constitutional limit on debt prevented the municipality

---

[1]The Hinsdale Sanitary District, organized in 1926, formerly was in Cook and Du Page Counties, but the Cook County portion was disconnected in 1970. In 1956 the Cook County portion of the Hinsdale Sanitary District was mistakenly annexed to the MSDGC. Since the disposal plant and part of the Hinsdale municipal sewerage lines are in Cook County, some arrangement had to be made. The Hinsdale Sanitary District processes sewage in the Cook County portion of the municipality of Hinsdale, even though this territory lies in the MSDGC. In return, the MSDGC processes sewage from the Du Page County portion of the municipality of Hinsdale, in an amount equal to the Cook County sewage treated by the Hinsdale Sanitary District.

[2]W. M. Olson, "The Value of Sanitary Districts," The American City, XXVII (December, 1922), pp. 557-63.

[3]Illinois, Annotated Statutes, ch. 42, sec. 306-311.

[4]Bloom Township Sanitary District, Forty-First Annual Operation Report, May 1, 1968 to April 30, 1969, by J. Edward Meers (Chicago Heights, Ill.: Bloom Township Sanitary District, n.d.), p. 7.

from borrowing funds for additional construction. In order to circumvent this restriction, a sanitary district, coextensive with Chicago Heights, was created at an election on June 4, 1928. The municipal sewage treatment plant was transferred to the district, and the district was able to borrow the necessary funds because its debt was legally distinct from the debt of Chicago Heights. In subsequent years, the district modified its original raison d'être by becoming a regional sanitary district. In 1947 the newly created municipality of Park Forest joined the district, and in 1951 the village of South Chicago Heights came into the district.

The district has no interconnections with the MSDGC, although in 1956 its territory became contiguous to the MSDGC. The district is now considering expansion in Will County, an option not available to the MSDGC. The Northeastern Illinois Planning Commission recommends that the Cook County portion of the district remain as it is but that considerable expansion should take place in Will County.[1] The district has its own staff for managing the sewage disposal plant; however, engineering details involved in expansion are carried out by private firms.

River Conservancy Districts

The Addison Creek Conservancy District, formed at an election in 1956, is the only district of this type in Cook County (Figure 12). Although it is a rural-type district, its territory lies entirely within the municipal limits of Northlake and Stone Park. The main function of the district is to control flooding and pollution within Northlake. Channel improvements and the construction of retaining structures have done much to alleviate the flooding problem.[2]

State, county, and area-wide district governments had done little to alleviate the flooding problem in Northlake, necessitating some kind of action on the part of local residents. The district includes only the middle portion of Addison Creek and cannot manage the whole watershed. Moreover, the headwaters of the stream are in Du Page County, placing the headwaters area outside the jurisdiction of the MSDGC.

In 1925 the Legislature provided for the organization of river conservancy

---

[1] Northeastern Illinois Planning Commission, Regional Wastewater Plan: An Element of the Comprehensive General Plan for Northeastern Illinois (Chicago: Northeastern Illinois Planning Commission, 1972), p. 19.

[2] This district has no office of its own. Information was obtained from employees of the municipality of Northlake, August 3, 1971.

districts. The original purposes of such districts were to unify the control of
river systems, to promote sanitation, to prevent stream pollution, to develop
and maintain a safe and adequate water supply, to aid navigation, and to protect
fish life. A 1951 amendment to the original act gave these districts the additional powers to control and prevent flooding, to reclaim wet and overflowed
lands, to develop irrigation, to collect and dispose of sewage, and to provide
for forests, wildlife areas, parks, and recreational areas.

The five-member board of trustees of the district is empowered to levy
taxes and special assessments, to issue bonds with voter approval, to exercise
the right of eminent domain, to construct dams, ditches, and pumping stations,
and to prevent the pollution of the water supply.[1] This wide range of powers is
unusual for a Special District in Illinois, for the Legislature traditionally has
been reluctant to grant more than a minimum of powers to a Special District. In
practice, the powers overlap those granted to other governments, and the Addison Creek Conservancy District applies only a very small number of its authorized powers.

Library Districts

There are fourteen library districts in or partly in Cook County, encompassing an area of 105 square miles. These districts are relatively recent in
Cook County history, the first ones dating from the late 1950's. For the most
part, library districts are located in suburban areas that have experienced a
sudden growth in population. The sudden explosion of population exerted great
demands on the existing political system to supply library services, and the creation of library districts were convenient ways of meeting demands with minimal
alterations in municipal government.

Four factors, acting singly or in combination, seem to account for the
formation of library districts: (1) a new municipality had more pressing needs
for its funds and could not afford to build a library; (2) an older municipality
might have been flooded by new residents who wanted library service, a service
that the existing power structure of the municipality was not willing to provide;
(3) people in an unincorporated area might have wanted a library but did not
desire municipal government; (4) the area flexibility of the district permitted
convenient areal arrangements to be made.

There are ten library districts which serve predominantly a single munic-

---

[1]Illinois, Annotated Statutes, ch. 42, sec. 383-410.

ipality. These are Streamwood, Wheeling, Niles, Roselle, Bensenville, Franklin Park, Bridgeview, South Stickney (serves Burbank), Bedford Park, and
Acorn (serves Oak Forest) (Figure 13). Three districts serve groups of municipalities. The Barrington Library District serves the municipalities of Barrington, Barrington Hills, and South Barrington. The Green Hills Library District
serves Hickory Hills and Palos Hills. The Forest View Library District serves
Forest View and Stickney. Only one district--Prospect Heights Library District
--serves primarily unincorporated territory.

Legal Aspects. --The first act for the creation of library districts was
passed in 1943.[1] The legislation was repealed in an act of 1957, which in turn
was modified by an act in 1967. Under the new law, the district board is empowered to levy a district tax and borrow money. It is the duty of the board to
establish, equip, and maintain a library or to contract with an existing library
for supplying library services to the district inhabitants.

The vote within municipalities and in unincorporated areas is counted
separately, and a favorable majority in each is required for the establishment
of a district. A municipality or township having its own tax-supported library
may, if it so desires, have its territory excluded from the library district. A
six-member board of commissioners elected for staggered six-year terms runs
the affairs of the district.[2]

Case Study: Niles Library District. --The Niles Library District was
organized at an election on April 18, 1959.[3] The municipality of Niles (incorporated in 1899) had no library of its own. The sudden growth in population
(from 3,587 in 1950 to 20,393 in 1960) led to a demand for library services by
residents who were used to them in Chicago and other suburbs. The municipal
government did not respond, leading to the rise of an interest group which was
successful in marshalling sufficient support to establish a library district nearly
coextensive with the municipality of Niles.

The library was first located in a park fieldhouse and then moved to a
converted store building. In 1964 land was purchased for a new library building,

[1]Laws of Illinois (1943), Vol. 1, p. 859.

[2]Illinois, Annotated Statutes, ch. 81, sec. 1001-1 to sec. 1008-2.

[3]Information on the Niles Library District obtained from an interview with
the Library Administrator, August 11, 1972.

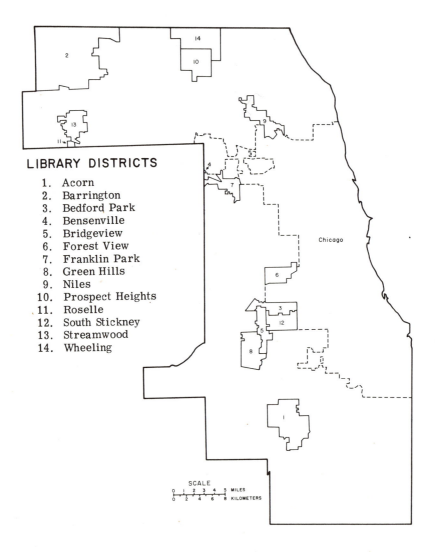

LIBRARY DISTRICTS

1. Acorn
2. Barrington
3. Bedford Park
4. Bensenville
5. Bridgeview
6. Forest View
7. Franklin Park
8. Green Hills
9. Niles
10. Prospect Heights
11. Roselle
12. South Stickney
13. Streamwood
14. Wheeling

Fig. 13.--Library Districts in Cook County

and in the same year a bond issue for $575, 000 was passed. This bond issue was supplemented by a $212, 000 federal grant, and numerous gifts were received from private sources. The library building was opened to the public in September, 1966 and now serves about 31, 000 people.

The library has 65, 000 books, a reference service, audio-visual services, and bookmobile service. The district is a member of the North Suburban Library System, a group of thirty-one member libraries in portions of Cook, Lake, and Kane Counties. Library card holders in any member library can borrow via interlibrary loan from the 2, 000, 000 volumes in the system libraries. In addition, card holders have access to the 3, 000, 000 books in the Chicago Public Library through an agreement.[1]

Mass Transit Districts

Introduction. --Mass transit districts are the most recent type of Special District to be formed in Cook County. The increased use of the automobile and accompanying decline in ridership of buses, railroads, and rapid transit lines have created many problems in traffic congestion, commuting, decline in commercial centers, and financial solvency for publicly and privately owned transportation facilities. In order to mitigate the stress on local governments caused by these problems, the Legislature in 1959 permitted the creation of local mass transit districts. These districts are empowered to coordinate the public transportation facilities within their areas. Moreover, since privately owned facilities cannot receive funds from State and Federal sources, the districts serve as convenient mechanisms for receiving grants from these higher levels of government.

Admittedly, these districts are a far cry from a bold, coordinated attack on the complex transportation problems besetting the Chicago area; nevertheless, the districts are politically feasible approaches.[2] They afford great local par-

---

[1]There are five groupings of libraries known as library systems in Cook County. These systems greatly increase the facilities of individual libraries by providing services to participating libraries. Instead of having a single unified county system, various parts of Cook County are functionally aligned with parts of adjacent counties. Chicago is considered a system by itself.

[2]In the spring of 1974 the Regional Transportation Authority (RTA) was approved for the counties of Cook, Du Page, Lake, McHenry, Kane, and Will. At the time of writing, this authority is still being organized. Ultimately, it will coordinate the activities of the mass transit districts and other governments concerned with transportation in the six-county area. The authority does not have the power to levy a property tax; therefore, it is excluded from the taxing Special Districts considered in this study.

ticipation in problem-solving and command strong support from municipal governments. Municipalities are able to exercise close control of these districts, insuring that no solutions are forthcoming that are inimical to local interests.

The areal flexibility of the district device permits a variety of governmental combinations to be formed in the six mass transit districts of Cook County (Figure 14). Three districts (Northwest Suburban, West Suburban, and Chicago South Suburban) include groups of municipalities astride rail lines extending outward from downtown Chicago. These districts own railroad coaches which are leased to the railroad companies serving the districts. The North Suburban Mass Transit District includes numerous municipalities and a township north of Chicago. This district leases railroad coaches to a railroad company, and it also runs its own bus lines.

The Des Plaines Mass Transit District was established in 1971 so that the village of Des Plaines could investigate the possibility of acquiring the local bus lines.[1] It was felt that the village itself could not do this because of possible legal difficulties, since the village receives money from the state gasoline tax.

The Chicago Urban Transportation District occupies nine square miles in the downtown area of Chicago. This district is discussed in more detail as a case study.

Legal Aspects. --A mass transit district may be formed in two ways: (1) one or more municipalities or one or more counties may combine, or (2) a "participating area" may be formed without regard to boundaries of other political units. The composition of the board of trustees varies with the number of political units that are contained within the district. If only one county or one municipality participates, there are three trustees. If there are more than one county or municipality, there is one trustee for each 50,000 people, provided each political unit has at least one trustee. The board of trustees may extend services beyond the territorial limits of the district with the permission of the corporate body of the area affected as long as they do not enter into direct competition with existing facilities. With referendum approval, the trustees may levy a tax, not to exceed .05 per cent, upon property within the district. They also have the power to invest inactive funds, borrow money, issue revenue bonds, and issue tax anticipation warrants.[2]

---

[1] Interview with the Mayor of Des Plaines, August 9, 1972.

[2] Illinois, Annotated Statutes, ch. 111-2/3, sec. 351-359.

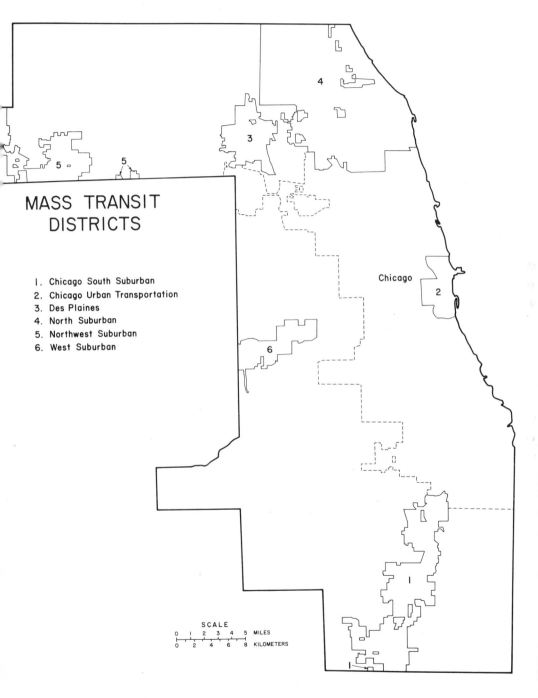

MASS TRANSIT
DISTRICTS

1. Chicago South Suburban
2. Chicago Urban Transportation
3. Des Plaines
4. North Suburban
5. Northwest Suburban
6. West Suburban

SCALE

Fig. 14.--Mass Transit Districts in Cook County

Case Study: Chicago South Suburban Mass Transit District. --The Chicago South Suburban Mass Transit District, created in 1967, was formed primarily to improve rail passenger service.[1] The thirty-three-square-mile district contains the municipalities of Riverdale, Dixmoor, Harvey, Hazel Crest, East Hazel Crest, Homewood, Flossmoor, Olympia Fields, Matteson, Park Forest, Richton Park, and Park Forest South (Figure 15). These municipalities are commuter suburbs that grew up along the Illinois Central Railroad, extending south from Chicago. The district levies no tax nor does it have any bonded indebtedness. The Board of Trustees consists of one trustee from each municipality, designated by the governing body of each municipality for a four-year term.

The district took advantage of the Federal Transportation Act of 1964 under which the district could receive funds for the purchase of new transportation equipment. Through the district device, the Illinois Central Railroad was able to obtain new rolling stock. The railroad provided one-third of the funds for new equipment by giving the funds to the district. The district then used these funds as its share for financing new equipment, and the Federal Government paid the remaining two-thirds. The district purchased 130 new railroad coaches and gave them to the railroad on a twenty-five year lease; the railroad maintains and operates the equipment.

The district has also acquired fifty-five buses which are leased for twelve years to the local bus company (South Suburban Safeway Lines, Inc.). Unlike the Illinois Central Railroad, however, the bus company did not have to put up any money for new equipment. The State of Illinois provided one-third of the cost of the new buses, and the Federal Government provided the remaining two-thirds.[2]

The district receives no money from the municipalities within its boundaries. Its operating funds are obtained from the Illinois Central Railroad. The trustees are paid twenty-five dollars per meeting, with a maximum of four meetings per month.

---

[1] Interview with Executive Director of the Chicago South Suburban Mass Transit District, August 12, 1974.

[2] In August, 1973 the funding ratio was changed from one-third local and two-thirds Federal to one-fifth local and four-fifths Federal. Of the one-fifth local share, the State of Illinois now pays 80 per cent and the local area 20 per cent. In some cases, the State may pay all of the local share. Source: Interview with officials of Illinois Department of Transportation, August 20, 1974.

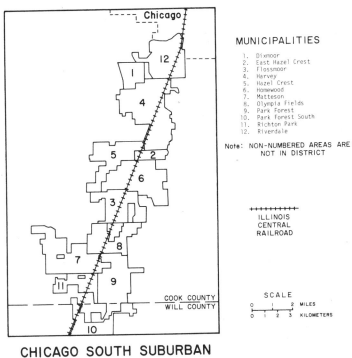

**CHICAGO SOUTH SUBURBAN
MASS TRANSIT DISTRICT**

MUNICIPALITIES

1. Dixmoor
2. East Hazel Crest
3. Flossmoor
4. Harvey
5. Hazel Crest
6. Homewood
7. Matteson
8. Olympia Fields
9. Park Forest
10. Park Forest South
11. Richton Park
12. Riverdale

Note: NON-NUMBERED AREAS ARE
NOT IN DISTRICT

ILLINOIS
CENTRAL
RAILROAD

SCALE

Fig. 15

Case Study: Chicago Urban Transportation District. --The nine-square-mile Chicago Urban Transportation District encompasses only the central business district of the city. Although established by a referendum in 1970, the district remained inactive until 1973 while its status was tested in the courts.[1] The district was the outgrowth of a planning study conducted in 1968 by the city of Chicago. The purposes of this plan were to improve the distribution of transit and commuter passengers, to remove the existing elevated structure, and to extend the rapid transit service. The district does not have a large staff of its own for research purposes; rather it relies heavily on private consulting firms for technical studies.

The district has a tremendous task in coordinating its activities with other governments and with public and private groups. It receives funds from the Federal and State Governments. It also must work with many agencies of the city of Chicago, the Chicago Transit Authority, the Cook County Department of Highways, the Chicago Area Transportation Study, the Northeastern Illinois Planning Commission, and the Regional Transportation Planning Board. The district also coordinates its activities with railroads, public utilities, and commercial and industrial interest groups.

Summary

The Special Districts serving municipalities have been very efficient devices by which the political system in Cook County has achieved homeostasis. For over a century, the political system has used Special Districts to satisfy public demands for new services or to deal with problems brought about by changes in the environment. At the same time, the districts have not seriously threatened the existing political structure.

The reasons for such Special District development are complex, but can be grouped into four general categories: (1) legal restrictions, (2) inaction of officials, (3) citizen attitudes, and (4) areal convenience.

State constitutional limitations on debt have prevented many municipalities from furnishing desired services. The Special District adds another layer of debt to the territory of the municipality but not to the municipality itself. The debt restriction appears to have been a major consideration in the formation of park districts and outlying sanitary districts.

---

[1]Information in this case study obtained from Chicago Urban Transportation District, Annual Report, 1973.

Although empowered to do so, certain facilities such as parks and libraries have not always been provided by municipal officials as fast as the public desired them. This output failure on the part of municipalities has been counteracted by the rise of interest groups which have found it easier to establish a district concerned with a particular function than to reform the municipal government.

Citizen attitudes concerning the desirability of maintaining district government separate from municipal government have made it easy for districts to be established and to command popular support. This attitude is strongly entrenched among Cook County citizens; and even though other factors that are important in the creation of districts may change, this attitude will probably guarantee the survival of the district form. As Willbern remarks:

> The special district is the most extreme manifestation of the almost universal desire for separateness and autonomy on the part of special constituency and clientele groups. Some of our sense of community is geographical, it is true, but increasing specialization of society and the organization and increasing strength of specialized interests lead some groups to feel stronger loyalty to a functional community or activity than to a geographical one. Even where the general purpose unit, the city or the county, operates a special function, the particular clientele or constituency involved with that function tends to demand as much autonomy as possible. Autonomy strengthens the lines of particularistic responsibility and helps to avoid lines of responsibility running through the general political system. [1]

Areal flexibility has also been a strong factor in the formation of some districts. As was noted in the discussion of park districts, a wide variety of governmental combinations is possible. Neighborhoods can have their own park districts inside a municipality; municipalities can join together for a park district; and pieces of unincorporated territory may be included within a park district that primarily serves a municipality. Municipalities can join with one another for certain services, but they still can retain their independence in providing other services. Mass transit districts are good examples of this type.

---

[1] York Willbern, The Withering Away of the City, Midland Books (Bloomington, Ind.: Indiana University Press, 1964), pp. 41-42.

CHAPTER VI

URBAN FRINGE DISTRICTS

In recent years the wide use of the automobile has made possible much
population growth in Cook County in areas that are transitional between urban
and rural. The growth of this area, known as the urban-rural fringe, or simply
urban fringe, has created a demand for services that has outstripped the capacity of the traditional county and township governmental structure to provide them.
Special Districts have provided urban-type services for residents who want the
amenities of urban life, but who do not wish to have a municipal government to
provide them.

Definition of the Urban Fringe

There is general agreement among users of the term that the fringe consists of a circumferential belt of land surrounding the central city, separating
the city from non-urban land uses.[1] Beyond this general consensus, however,
the definition is less precise. Some writers use the term fringe to refer to any
area surrounding a municipality, even if that municipality is a suburb of a larger
city. As Murphy says, "To a lesser degree the phenomenon is repeated around
satellites and suburbs."[2]

The Census Bureau uses the term urban fringe to refer to the urbanized
area minus the central city.[3] According to this definition, suburbs that are

---

[1] For a collection of papers on some of the problems of the urban fringe,
see Jean Gottmann and Robert A. Harper (eds.), Metropolis on the Move: Geographers Look at Urban Sprawl (New York: John Wiley & Sons, 1967).

[2] Murphy, The American City, p. 43.

[3] The term "urbanized area" is used by the Census Bureau to refer to the
built-up area of a city, regardless of the political organization. In general, the
urbanized area is determined on the basis of population density and land use.
For a fuller discussion, see Murphy, The American City, pp. 24-30.

incorporated would be part of the fringe. In common practice, the term is loosely used to refer to all of the built-up area surrounding a central city whether incorporated or not. The term also is used for areas that may extend beyond the urbanized area as determined by the Census Bureau.

Urban fringe, as used in this study, will refer to the unincorporated areas consisting of substantially built-up land located near the border but outside the legal boundaries of existing municipalities. All of Cook County outside municipalities fits into this definition reasonably well. Little agricultural land remains in the county, and no point in the county is more than four miles from a municipal boundary.

Nature of the Fringe

Urban fringe growth has been made possible because of the greater freedom of movement resulting from the increased supply of automobiles and major improvements in expressways and access roads. People can live farther from their places of employment and satisfy their needs from outlying business centers and highway-oriented businesses.

People move to the fringe not only because of cheaper land, but also because they desire more space, air, sunshine, and a feeling of freedom. Many fringe dwellers, regardless of their economic status, want to get away from the noise, dirt, crime, and congestion of the city to neighborhoods which are quieter, cleaner, safer, and less crowded. The desire for physical escape from the city does not, however, necessarily mean a desire to leave city conveniences and attractions. Furthermore, the longing for openness is often not realized permanently, for many of these newly developed areas later become solidly built-up.

Although not often perceived by local residents, many urban fringes are actual or potential problem areas because their standards and services are frequently inferior to those maintained by the municipalities they border. Land uses are often incongruous mixtures of industrial, commercial, and residential properties. Incompatible land uses and deficient subdivision layouts are costly to rectify as an area becomes more thickly settled. Moreover, the deficiencies of these areas can adversely affect the health and safety of not only fringe residents but also inhabitants of the neighboring municipalities. Mistakes due to lack of planning become imbedded in the landscape and are costly to correct.[1]

---

[1] For an account of the conditions in unincorporated areas in Cook County three decades ago, see Eugene F. Christgau, "Unincorporated Communities in Cook County" (unpublished M.A. dissertation, Department of Sociology, University of Chicago, 1942).

One of the most important deficiencies of fringe areas is the absence or inadequacy of public services. Built-up areas, whether or not incorporated, need a minimum level of services in order to survive. Most urban fringe areas suffer significant deficiencies in this respect, although their residents usually manage to obtain minimal services from public or private sources. The most prevalent shortcomings are the inferior condition of sewage disposal, sanitation, and drainage. Other services that are ignored or only partially satisfied include street construction, paving, sidewalks, street lighting, water supply, fire protection, law enforcement, recreation, and public health.

Government in the Fringe

The fringe has many urban characteristics but lacks general-purpose urban government. Most services are provided by the county and township governments and by a host of Special Districts; a few services are provided by private sources and by municipalities outside their borders. Some of the services provided by Special Districts have been discussed in other chapters, but here we are interested in those Special Districts serving mainly the urban fringe, or simply fringe districts.

In Cook County the number of fringe districts has grown more rapidly than any other type of district. In 1939 there were only three Special Districts that lay entirely within unincorporated territory. After World War II, an estimated eighty districts had been formed in unincorporated areas, although in 1973 only sixteen of these districts still had no municipal territory within their boundaries.

Four factors stand out as contributing to the great growth of fringe districts in Cook County: (1) the lack of authorized powers for counties and townships, as well as the absence of a financing mechanism by which counties or townships can make improvements to be paid for only by those benefited; (2) the determined opposition by many fringe residents to placing their areas under municipal governments; (3) the practice of municipalities providing services to urban fringes; and (4) the availability of appropriate legislation for the formation of Special Districts.

Legal Restrictions. --The question may well be asked why the county and townships do not supply the services now rendered by Special Districts. Part of the answer is that neither the county nor the township is legally authorized to perform these services. The Illinois Legislature has been very reluctant to grant additional powers to townships and counties so that these units of government could respond to citizen demands. Instead, the Legislature has adopted the district device, which allows a single service to be performed for a small area. This piecemeal approach to satisfying the service needs solves the immediate problem, but it effectively prevents an orderly coordinated approach to government in the urban fringe. As Elazar comments:

> The effort to limit the exercise of governmental powers in each locality and to fragment those that are exercised has been another apsect of the transplantation of agrarian values into urban settings. The reasons for this are complex and the effort itself is by no means consistent. Particularly in the region of the cities of the prairie, but throughout the nation as well, there is a hesitancy about strengthening local government in the provision of amenities for fear that the addition of more local services will increase the urban character of the environment. [1]

In addition to the lack of authorization of powers, another stumbling block to the provision of services by counties and townships was found in the Illinois Constitution of 1870, which did not give the powers of special assessment and establishing differential taxing areas to counties and townships.

Special assessment powers were given only to cities, towns, villages, and drainage districts in the 1870 Constitution. Counties and townships could not make special assessments in their territories. As a result, if part of a township or county wanted a special service, and this part was not within the jurisdiction of a governmental form that had the power of special assessment, the service could not be provided. [2]

Another legal limitation on local government was that under the Illinois Constitution of 1870, differential taxing areas were not permitted. The Constitution stated that all areas within the jurisdiction of a local governmental form had to be taxed equally. In the words of the Constitution:

> For all other corporate purposes, all municipal corporations may be vested with authority to assess and collect taxes; but such taxes shall be uniform, in respect to persons and property, within the jurisdiction of the body imposing the same. [3]

In another section of the Constitution, the restriction is stated in slightly different terms:

---

[1] Daniel J. Elazar, Cities of the Prairie: The Metropolitan Frontier and American Politics (New York: Basic Books, 1970), p. 49.

[2] Illinois, Sixth Illinois Constitutional Convention, 1969-1970, Report of Committee on Local Government (Springfield, Ill., n.d.), p. 85.

[3] Illinois, Constitution (1870), art. IX, sec. 9.

The General Assembly . . . shall require that all the taxable property
within the limits of municipal corporations shall be taxed for the payment
of debts contracted under the authority of the law, such taxes to be uniform
in respect to persons and property within the jurisdiction of the body impos-
ing the same.[1]

The legal inability of townships and Cook County to establish special tax
rates for small areas within their boundaries and to levy special assessments
probably has encouraged numerous Special Districts to be created. The townships and county did not need special services in all of their areas, and the district device permitted these services to be utilized and paid for by the residents
most concerned.

There has been much criticism of the great number of Special Districts
that have been formed in Illinois to circumvent the constitutional restrictions on
powers of the county. The new 1970 Constitution grants power to the county (but
not township) to levy special assessments and create special taxing units. As
noted in the new Constitution:

The General Assembly may not deny or limit the power of home rule units
(1) to make local improvements by special assessment and to exercise this
power jointly with other counties and municipalities, and other classes of
units of local government having that power on the effective date of this Constitution unless that power is subsequently denied by law or (2) to levy or
impose additional taxes upon areas within their boundaries in the manner
provided by law for the provision of special services to those areas and for
the payment of debt incurred to provide those special services.[2]

Since Cook County is a home rule county, it now has the legal authorization to make special assessments and set up differential taxing areas.[3] This
may be considered to be a desirable reform, but it is not likely to lead to any
substantial change in district organization within the next few years. There
first must be the authorization of additional powers from the Legislature. Such
authorization would depend on the articulation of demands from citizens for
those powers to be granted, and no ground swell of public opinion has been
discerned.

Even if Cook County had additional powers, there would have to ensue a
long political process to create sub-areas of the county that would have additional taxes for additional services. The possible opposition of municipalities

---

[1] Illinois, Constitution (1870), art. I, sec. 10.

[2] Illinois, Constitution (1970), art. VII, sec. 6(1).

[3] There are no differential taxing areas in Cook County at the present time,
nor is the county levying any special assessments. This information obtained
from officials in the office of the Cook County Clerk, August 14, 1973.

and existing districts would have to be overcome. Also, the interests of local groups in the proposed sub-areas would have to be satisfied. Ultimately, some sort of dependent district device might be worked out, in which the county would perform the service and local control would be maintained in the sub-area.

Resistance to Urban Government. --A major element in the growth of fringe districts is the belief of many urban fringe dwellers that they need, or are at least willing to finance, only a few public services. Despite the heavy urbanization ultimately experienced in most urban fringes, the belief in a rural setting remains for a long time. Even when their areas are densely settled, urban fringe residents are likely to resist the traditional processes of municipal incorporation or annexation.

The very existence of fringe districts often is an obstacle to municipal government. Once an area has solved its basic problems of water supply, sewage disposal, and fire protection through Special Districts, it is likely to ignore the less obvious advantages of municipal government. Even when incorporation does take place, preference for the district form of government still persists. In many parts of Cook County, in which municipal government has recently been established, one finds both municipal and erstwhile "fringe district" government operating side by side.

The utilization of annexation is restricted by the requirements of fringe initiation and approval, as well as municipal approval; these requirements have greatly curtailed the use of this method. No matter how obviously the fringe territory is part of the adjoining municipality in terms of social and economic realities, the judgment of the people living in the unincorporated fringe is the controlling factor. This policy has been slightly modified in recent years by the Legislature. On its own initiative, a municipality may annex a parcel of property of less than sixty acres, provided the parcel is completely surrounded by one or more municipalities. [1]

Public acceptance of fringe districts is based on the assumption that creating a Special District may be the only alternative to incorporation or annexation and that it is cheaper than municipal government. The same assumption was found to exist in the San Francisco area:

> The creation of a district is often believed to be a less expensive way of obtaining service than incorporation as a municipality or by annexation to an existing city. Thus there is a good deal less resistance to the creation

---

[1] Illinois, Annotated Statutes, ch. 24, sec. 7-1-13.

of a special district than to municipal incorporation or annexation. [1]

Whether the costs are actually lower in the fringe than in the municipality is a controversial question. According to Bollens, the fringe dwellers are apparently not aware of the fact that the gradual accretion of Special Districts eventually results in uneconomical government. He maintains that the comparative cost of the same amount of service in the fringe and in the nearby city generally favors the city. Furthermore, fringe costs based on total taxes, higher insurance rates, and higher utility charges sometimes exceed comparable outlays in the neighboring city, even when the city provides more and better services. [2]

Provision of Services. --The ease by which Special Districts may provide services also has encouraged their growth. If such services were not so readily available, the fringe area would have to turn to municipal annexation or incorporation. One of the reasons for Chicago's great growth in the nineteenth century was the necessity of relying on Chicago's municipal government to provide urban services. Large quantities of rural land were brought into the city in the 1880's and 1890's, and the municipal government could make orderly plans for growth. The great cost of providing services made it essential to have available the pooled resources of many people.

Today, the great increase in wealth has not made such large-scale operations necessary. Most fire protection districts now have their own equipment, each one providing protection for a few thousand people. Collector sanitary districts can rely on the MSDGC to dispose of sewage, and numerous small sewerage systems can be financed out of current revenues. Fringe areas also can rely on nearby municipalities for certain services such as police, fire protection, water supply, and recreation. These services can be provided directly to fringe residents by contract, through a schedule of fees, or through the district mechanism. Both the county and township provide street maintenance and snow removal under the guise of road maintenance. This piecemeal approach to satisfy service needs is a fringe substitute for municipal incorporation or annexation.

Enabling Legislation. --The growth of urban fringe districts has been facilitated by the ease with which enabling legislation has been passed by the

---

[1] Scott and Corzine, Special Districts in the San Francisco Bay Area, p. 5.

[2] Bollens, Special District Governments, pp. 114-15.

Legislature. For decades the Legislature has passed enabling legislation for Special Districts rather than attempting to make the traditional governmental structure more responsive. Without permissive laws, the urban fringe would be compelled to turn to incorporation or annexation. A large amount of Special District legislation that is usable by urban fringes has been enacted by the Legislature. As was shown in previous chapters, many districts serve both municipalities and fringe areas.

Although all of the possible types of districts are not represented in Cook County, legislation enables districts to be formed for such traditional urban functions as parks, fire protection, public health, libraries, street lighting, sanitation, water supply, hospitals, and airports. These districts are not legally restricted to municipalities, and many of them have fringe territory within their borders.

Characteristics of Fringe Districts

Three types of Special Districts have been designated as fringe districts: fire protection, collector sanitary, and street light. More than half of the territory of these districts lies outside municipalities (Table 13); yet they are not large enough to be considered area-wide in scope.

TABLE 13

AREAL CHARACTERISTICS OF FRINGE
DISTRICTS IN COOK COUNTY

Type	Number	Area (Sq. Mi.)	Per Cent of District Area Covered by Municipalities
Fire Protection	45	289	41
Collector Sanitary	25	21	48
Street Light	1	0[a]	0

[a]Less than one-half square mile.

These districts, for the most part, are located in the newly urbanized parts of Cook County, particularly in the northwestern townships. The townships in which suburban development was extensive before World War II, such as Proviso, New Trier, and Thornton, have few of these districts. When the latter townships were urbanized, auto transportation was less important than it is today, resulting in a more compact type of settlement. Moreover, the lack of enabling legislation for many types of Special Districts was not enacted until recent decades.

A fringe district may be organized at various stages in the development of the urban fringe, but once established, the district does not usually expand in size as the fringe area grows. Residents of the newer settled portions of the fringe desire separate and highly localized control of a function. Other factors in the creation of a new district may be the dislike of officials of the existing district, or the newer residents may not want to share the outstanding debt of an established district.

Some writers consider fringe districts as a stage in a sequence of development.[1] The first stage consists of a scattered population existing in an essentially rural area, with a minimum of township and county services. The second stage is marked by a substantial increase in population and by a marked increase in demand for urban services such as water, sewage disposal, and fire protection. It is in this stage that Special Districts are created to provide critical services, since the county or township governments cannot or do not wish to supply the needed services. The third stage begins when the residents of the fringe area realize that they are in fact an urban community, and that they require urban services. Once this realization is reached and the requisite political forces can be mobilized, municipal incorporation or annexation to an existing municipality follows.

This sequence of governments portrayed as stages of development has many parallels in Cook County. Many of the fringe districts of today are in areas that formerly had only township, county, and school district government. Moreover, some municipal areas of the county formerly had fringe district government. Certainly the process has occurred in many parts of Cook County, but it is by no means a necessary or inevitable process. Many municipalities have expanded their boundaries rapidly, obviating the need for fringe districts. On the other hand, some unincorporated areas are perfectly satisfied with fringe district government, and there is no discernible sentiment for municipal government.

---

[1]Victor Roterus and I. Harding Hughes, Jr., "Governmental Problems of Fringe Areas," Public Management, XXX (1948), 94-97. Also Victor C. Hobday, "Should Cities Provide Services to Suburbs or Extend City Limits?" Illinois Municipal Review, XXX (February, 1951), 24.

The greatest confusion occurs where incorporation or annexation takes place after fringe districts have already been established. Conversations with officials have revealed that only about a dozen districts have been dissolved and their functions taken over by municipalities. For example, Oak Lawn has absorbed several small fire protection districts that once were in unincorporated territory adjacent to the city but the territory now has been annexed to the city. In other cases, Chicago Ridge and Rolling Meadows were formerly coextensive with fire protection districts, but the district governments were transformed into municipal fire departments.

A much more common occurrence has been the creation of a new relationship between the municipality and Special District. Most Special Districts do not disappear, even though the conditions that brought about their creation have changed. Just like other areal political forms, the district often acquires a new raison d'être, develops its own interest groups, and builds up a loyal following, all combining to ensure the perpetuation of the district. Moreover, the complex legal and fiscal arrangements surrounding Special Districts are difficult to change. For example, bonds that cover the outstanding indebtedness of the district are payable over a long period of time, and the municipality might not be willing to accept new financial obligations. In most cases, the district merely assumes a municipal function, although it is legally distinct from the municipal government.

Fire Protection Districts

Introduction

Cook County has forty-five fire protection districts within or partly within its boundaries (Figure 16). These districts encompass an area of 289 square miles, of which about three-fifths lies in areas outside municipal boundaries.

A fire protection district act was first passed by the Legislature in 1927. Under the terms of this act, fifty or more voters could petition the county judge to hold an election for the establishment of a fire protection district. The petitioners would suggest a boundary for the proposed district, but the judge could alter the proposed boundary, if he wished to. A favorable majority of the votes cast at an election was sufficient for establishing the district. No restrictions in the act were placed on the presence or absence of incorporated territory.[1]

A few organizations, similar to fire protection districts, existed many

---

[1] Laws of Illinois (1927), pp. 530-36.

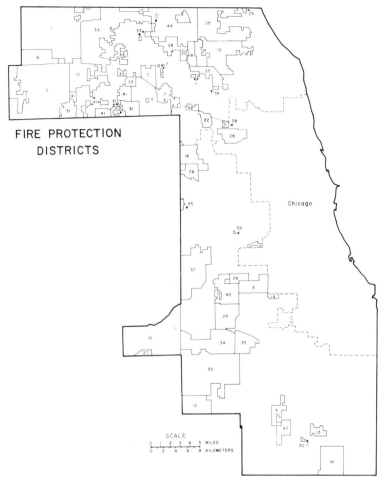

FIRE PROTECTION
DISTRICTS

1. Barrington Countryside	16. Leyden	31. Ontarioville
2. Bartlett Countryside	17. Long Grove Rural	32. Orland
3. Burbank Manor	18. Miller Woods	33. Palatine Rural
4. Central Stickney	19. Mokena	34. Palos
5. Country Club Hills	20. Mount Prospect	35. Palos Heights
6. East Dundee	21. North Arlington	36. Park Ridge Manor
7. Elk Grove Rural	22. North Leyden	37. Pleasantview
8. Forest River	23. North Maine	38. Prospect Heights
9. Forest View	24. North Palos	39. Riverside Lawn
10. Glenview Rural	25. Northbrook Rural	40. Roberts Park
11. Hawthorne Rand	26. Northlake	41. Roselle
12. Hoffman Estates	27. Northwest Homer	42. South Maine
13. Holbrook	28. Norwood Park	43. Sunnycrest
14. Hometown	29. Nottingham Park	44. Wheeling Rural
15. Lemont	30. Olympia Gardens	45. Yorkfield

Fig. 16. --Fire Protection Districts in Cook County

years before 1927 as corporations, cooperatives, or units receiving protection by contract from a city or village fire department. Existing principally by donations, the cooperative often found itself short of funds when new equipment was needed. One of the reasons for the fire protection district act was to assure funds to buy equipment and maintain it. Also, it seemed desirable for everyone in the district to pay part of the cost rather than to depend on donations from a few--the few being the people who had fires and thus were least able to contribute. [1]

The first fire protection district in Cook County was formed about 1940. [2] After World War II, there was a great out-migration of population from Chicago to housing developments and scattered housing in unincorporated areas. In many cases, these newly arrived people did not want to become part of a municipality, and yet they wanted fire protection. As a result, many districts were created in essentially rural territory, but which was fast acquiring some urban characteristics. An older municipality, particularly if small, occasionally would be included within the district (such as Bartlett).

Operations of Fire Protection Districts

A fire protection district may be one of three general types: (1) it may have its own fire-fighting equipment in its own fire station; (2) it may own or share equipment, which is housed in the fire station of a municipality (which may or may not be within the district); or (3) it may contract with a municipality for fire protection. If it has its own fire station, it may be manned by volunteer or full-time personnel.

Through a long, time-consuming series of queries of officials, a survey was made of the forty-five fire protection districts in Cook County. Thirty-one had their own equipment in their own fire stations; six had their own or shared equipment housed in a municipal fire station; five obtained protection by contract from a municipality; and three could not be determined.

Because of the expense, most fire protection districts are unable to

---

[1] N. P. G. Krausz, "Fire Protection Districts in Illinois," Quarterly of the National Fire Protection Association, XLIV (July, 1950), 178-91.

[2] Information obtained from the Illinois Inspection and Rating Bureau, which is the official fire insurance rating bureau in Illinois. As additional evidence, a 1939 atlas shows no fire protection districts in Cook County. Source: Illinois, Tax Commission, Survey of Local Finance in Illinois, Vol. I: Atlas of Illinois Taxing Units (Springfield, Ill., 1939), pp. 47-52.

employ full-time firemen. As a result, they are manned by volunteers, most of whom serve without pay. Various arrangements are made to report a fire at any time of day or night, and the absence of full-time personnel is not a serious handicap. For example, a district may maintain an answering service, in which the operator pushes a special button that notifies everyone of the need for fire duty. In those districts containing old municipalities, the firehouse is centrally located; and people living near the fire station are available for call day or night.

Fire protection districts and municipal fire departments have agreements with one another for mutual aid. Some municipal fire departments will also provide assistance, even though no written agreement exists. Although some of the legal obligations are not clearly spelled out, the system works remarkably well on a basis of cooperation. For example, if district equipment is busy fighting a fire, other fire departments will provide protection in other parts of the district. In emergencies, the suburban municipalities and fire protection districts have an unwritten agreement with Chicago to provide or receive aid.

Cook County has a dense network of all-weather roads which facilitates rapid access to all parts of the fire protection district. There is a practical limit to the area that may be served by a single fire station. For fire insurance purposes, no point in the district should be more than ten miles from a fire station. Fire experience has shown that the time lost in covering a distance of more than ten miles generally results in total fire loss. Most fire protection districts are small enough to easily come within the prescribed limit.

Fire Insurance Rating

The Illinois Inspection and Rating Bureau is the official fire insurance rate-making body in Illinois. Many factors are used by the bureau in determining fire insurance rates, including occupancy, contents of buildings, structural materials, fire hazards, and quality of protection. The primary concern here is the quality of protection provided by fire protection districts. The bureau does not release detailed information, although an over-all assessment can be made.

Areas are divided into ten classes according to the degree of fire protection, with Class 1 providing the best protection and Class 10 the least protection. Areas with no fire department, not served by another fire department under contract, nor in a fire protection district are in Class 10. A fire protection district with minimal protection, such as a volunteer fire department and no hydrants, would be in Class 9. A district can move up the scale by having its own equipment, maintaining full-time personnel, and having access to fire hydrants with dependable water pressure. Chicago, with an elaborate fire-fighting system, is in Class 2. Most districts are in Class 9, and a few go as high as Class 6. All other things being equal, moving up in class results in lower fire insurance rates.

Fire Protection Districts and Municipalities

Fire protection districts are created predominantly in unincorporated territory, although a few have included municipalities initially. As urbanization continued, many parts of the districts became subject to municipal government. Three broad categories of responses to this change in government have occurred: (1) The district continues to provide fire protection; (2) The district is dissolved; and (3) The incorporated part of the district is detached.

District Continues to Furnish Fire Protection. --Fire protection is provided by fire protection districts to twenty-seven municipalities in Cook County. For the most part, municipal incorporation took place after the district was formed, and the district continued to provide fire protection as before. The municipality simply does not have a fire department. In five known cases, an older small municipality was included within the district when formed. These municipalities were Bartlett, Lemont, Orland Park, Justice, and Palos Park.

In some cases, the district arrangement permits favorable areal combinations. The Norwood Park Fire Protection District serves the villages of Norridge and Harwood Heights and some unincorporated territory; the district territory comprises two enclaves within Chicago. The Lemont Fire Protection District includes the village of Lemont and considerable surrounding territory that is unincorporated. The Pleasantview Fire Protection District provides protection for the small municipalities of Burr Ridge, Countryside, and Indianhead Park. Interestingly enough, it is possible for a municipality to lie in more than one fire protection district.

District Is Dissolved. --Two situations may lead to the dissolving of a fire protection district. The more common situation is where a small district was formed on the edge of an older large municipality that already had a fire department, and the district territory was annexed to the municipality. The less common situation is where a municipality was formed inside a fire protection district, and the district facilities were taken over by the municipal government.

No statutory provisions exist for the dissolving of a fire protection district on the initiative of the district. However, if a municipality annexes the entire district, or is organized within the entire territory of the district, the municipality on its own initiative may dissolve the district.[1] If a municipality has more than 50 per cent of the district territory within its boundary, it may take over the district facilities on its own initiative, provided it furnishes fire protection to the remainder of the district outside the municipality.[2]

Several small districts near Oak Lawn have been taken over by the municipality. District facilities have been integrated with the municipal system, and the outstanding debt of the districts has been taken over by the municipality.[3] The municipalities of Chicago Ridge and Rolling Meadows have taken over completely the fire protection district facilities that formerly served the municipalities.

District Territory Is Detached. --Many large districts have been formed on the periphery of well-established older municipalities that already had fire departments. When parts of these districts were annexed to these older communities, the Legislature has decided that the municipality shall provide the fire protection and the newly annexed territory shall be disconnected from the district.[4] This disconnection from districts has been particularly prevalent in the northwestern townships of Cook County, where extensive rural fire protection districts have been formed. Since many of these districts have their equipment housed in municipal fire departments, the net effect of these changes has been to shift the use of the equipment owned by the district to that owned by the municipality. The personnel manning the equipment has remained substantially the same.

---

[1] Illinois, Annotated Statutes, ch. 127-1/2, sec. 38.2.

[2] Ibid., ch. 127-1/2, sec. 38.4.

[3] Conversation with officials in City Manager's Office, City of Oak Lawn, August 17, 1973.

[4] Illinois, Annotated Statutes, ch. 127-1/2, sec. 38.3.

Legal Structure

A fire protection district is governed by a board of trustees, consisting of three members who hold office for three years. Not more than one trustee may be from a city or village in the district unless that city or village has over 50 per cent of the population of the district. The board of trustees may appoint a fire chief and such firemen as it considers necessary. The board may also prescribe the duties and fix the compensation of all officers and employees of the district. [1]

The trustees are appointed in a complex manner. Formerly, they were appointed by the Circuit Court, but after 1971 this procedure was changed. If the district is coextensive with a township, the trustees are appointed by the township supervisors. If the district is within a municipality, then the trustees are appointed by the chief municipal officer. If the district is in unincorporated territory and entirely within one county, then the trustees are appointed by the chief executive officer of the County Board. If the district is in unincorporated territory and in more than one county, then the State senator and State representatives from that area choose the trustees. [2]

Case Study: Barrington Countryside
Fire Protection District

This district well illustrates the great functional utility of fire protection districts in the space-polity of Cook County. The district provides a vital service that cannot be provided legally by township or county government; it forms a very convenient arrangement for the district residents; and it poses no threat to municipal, township, or county government.

The district, organized about 1950, lies in the extreme northwestern part of Cook County and extends into McHenry and Lake Counties (Figure 17). The district is larger than most fire protection districts, having an area of 47 square miles (30 square miles in Cook County). The district primarily serves the high-income, low-density suburbs of Barrington Hills and South Barrington, although parts of the district extend into the municipalities of Lake Barrington, Inverness, and Hoffman Estates. The district device permits a large, reasonably compact service area to be formed within a multiplicity of jurisdictions of

[1] Ibid., ch. 127-1/2, sec. 21-46.

[2] Conversation with the Secretary of the Illinois Association of Fire Protection Districts, August 8, 1972.

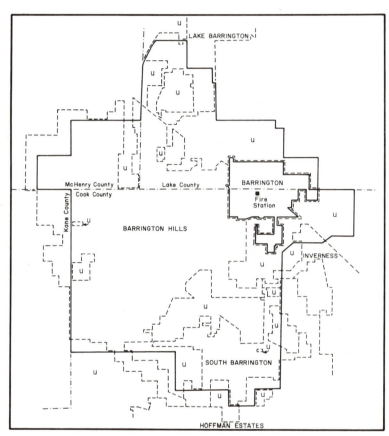

BARRINGTON COUNTRYSIDE
FIRE PROTECTION DISTRICT

--- MUNICIPAL BOUNDARY
—— DISTRICT BOUNDARY
U = UNINCORPORATED

SCALE

Fig. 17

general-purpose government; yet the district does not form a rival or threat to the power structures of any of them.

The district has no fire station or fire-fighting personnel of its own. It arranges for fire protection from the village of Barrington which has a volunteer company. The fire station in the village houses equipment belonging both to the district and to the village. By having two sets of equipment, additional protection is afforded in case of more than one fire at the same time.[1] This arrangement is very convenient for the residents of Barrington Hills which has no commercial center, and the assembling of volunteer firemen would be difficult.

Collector Sanitary Districts

Characteristics. --In addition to the sanitary districts of the 1889 type and 1917 type described in previous chapters, there are twenty-five sanitary districts in Cook County of the 1936 type--called collector sanitary districts (Figure 18). These districts, lying within the territory of the MSDGC, own their own sewerage lines and discharge their effluent into interceptor sewers owned by the MSDGC; the effluent is then treated by MSDGC disposal plants.[2] Some of these districts also provide a piped water supply to district residents.

When a fringe area is first developed, the houses are far apart, and each has its own septic tank and well. As population density increases, the soil is unable to absorb the effluent, and pollution of the water supply results. The creation of the sanitary district eliminates the health hazard and still retains the rural setting. If the demand for municipal services continues to increase, the area may eventually incorporate. If incorporation or annexation occurs, the functions of the district may be assumed by the municipality or the district may still continue to operate.

The total area of the twenty-five sanitary districts encompasses only twenty-one square miles. Most of the districts are very small, and the largest contains only 4.6 square miles. In nineteen of the districts, all or the majority of their territories lie outside municipalities. Six of the districts lie entirely within or mostly within municipalities, some of which have their own sewerage systems. Probably, these districts will be integrated into municipal systems

---

[1] Conversation with the Chief of the Barrington Fire Department, July 18, 1971.

[2] The Barrington Woods Sanitary District has a small treatment plant operated by the MSDGC. However, an interceptor sewer linking the district to one of the larger treatment plants of the MSDGC is under construction.

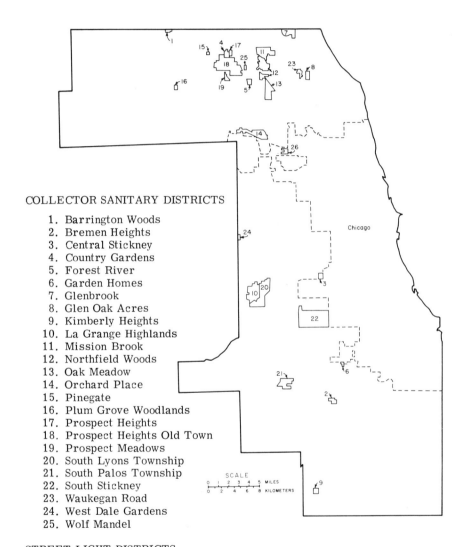

COLLECTOR SANITARY DISTRICTS

1. Barrington Woods
2. Bremen Heights
3. Central Stickney
4. Country Gardens
5. Forest River
6. Garden Homes
7. Glenbrook
8. Glen Oak Acres
9. Kimberly Heights
10. La Grange Highlands
11. Mission Brook
12. Northfield Woods
13. Oak Meadow
14. Orchard Place
15. Pinegate
16. Plum Grove Woodlands
17. Prospect Heights
18. Prospect Heights Old Town
19. Prospect Meadows
20. South Lyons Township
21. South Palos Township
22. South Stickney
23. Waukegan Road
24. West Dale Gardens
25. Wolf Mandel

STREET LIGHT DISTRICTS

26. Norwood Park

Fig. 18.--Collector Sanitary and Street Light Districts in Cook County

at some future date, but the process of integration is slow since the sewerage lines are already installed in a relatively fixed pattern. Moreover, the debt structure of the district is set up for many years. The municipalities of Burbank and Countryside do not have water or sewer systems; they are served completely by districts.

Legal Aspects. --In 1936 the Legislature approved an act allowing the formation of sanitary districts in unincorporated territory, provided the districts were entirely within a single county. This law does not prohibit the formation of sanitary districts, even if the territory is already within another sanitary district formed under laws of a different year. Since sanitary districts of the 1936 type are also within sanitary districts of the 1889 type (the MSDGC), taxes for both types are collected. Although not legally defined, in practice the 1936 type owns the sewerage lines (hence the term "collector"); and the 1889 type disposes of the sewage in treatment plants. The 1936 type of sanitary district may also acquire and operate waterworks with voter approval.[1]

Unlike many Special Districts in Illinois, sanitary districts of the 1936 type have procedures for dissolution. Fifty voters may petition the Circuit Court for an election to dissolve the district.[2] If part of a 1936-type sanitary district is annexed to a municipality, the district still continues to operate within the municipality. The entire sanitary district must be annexed to a municipality, or the district legally dissolved, for the municipality to take over its functions.[3]

The district operates under the guidance of three trustees, who are appointed for three-year terms. The trustees were formerly appointed by the Cook County Circuit Court, but since 1971 they are appointed by the President of the Cook County Board, with the approval of the remaining board members.[4]

Case Study: South Stickney Sanitary District. --The South Stickney Sanitary District was organized in 1951 in unincorporated territory adjacent to southwestern Chicago (Figure 19). This territory was considered as urban by the Census Bureau as far back as 1910, but the area managed to get along with minimal services until the great influx of population after World War II. Since that

---

[1]Illinois, Annotated Statutes, ch. 42, sec. 412, 443b.

[2]Ibid., ch. 42, sec. 444.   [3]Ibid., ch. 42, sec. 447.1.

[4]Ibid., ch. 42, sec. 414.

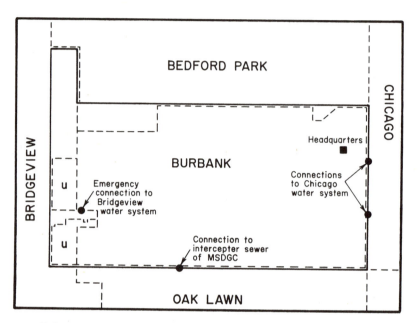

SOUTH STICKNEY SANITARY DISTRICT

u = UNINCORPORATED
— — — MUNICIPAL BOUNDARY
———— SOUTH STICKNEY SANITARY
DISTRICT BOUNDARY

SCALE
0 .1 .2 .3 .4 .5 MILE
0 .2 .4 .6 .8 KILOMETER

Fig. 19

time, the area has acquired additional services from districts for sanitation, public health, library services, parks, and fire protection. Some of the area was annexed to the villages of Bridgeview and Bedford Park, but the greater part of the area was incorporated as the city of Burbank in 1970.

By taking advantage of the 1936 act, the district could be formed within the jurisdiction of the MSDGC. The district owns and operates a sewerage system that collects liquid wastes and transports the effluent to an interceptor sewer of the MSDGC, which then conducts the effluent to its Southwest Disposal Plant. The district makes a minimum charge of $1.00 per month for a sewerage connection.[1]

The district also has its own water distribution system. It receives treated water from the Chicago municipal water system, but the district has its own water tank to maintain adequate pressure. Repairs to the water mains and sewer pipes are handled by private contractors. The district also has an emergency connection to the municipal water system of Bridgeview.

The Sanitary District Act of 1889 requires that a municipality taking water from Lake Michigan must sell water to a customer on its borders for the same rate as it charges its own customers. The district buys water from Chicago at twenty-seven cents per thousand gallons. It then sells the water at $1.20 per thousand gallons for the first 2,500 gallons and at $0.50 per thousand gallons for all amounts exceeding 2,500 gallons. Both municipalities and sanitary districts near Chicago make a considerable profit by selling water to their own citizens or to other suburbs at a higher rate than they pay Chicago.[2]

The district is probably only a transitional form between township government and municipal government. When the bonds for the district's indebtedness are paid up, the district will probably be dissolved and its functions assumed by the city of Burbank. The Bridgeview and Bedford Park areas within the district then would be served on a contract basis by the Burbank government.[3]

---

[1] Factual information obtained from an interview with the Superintendent of the South Stickney Sanitary District, August 2, 1970.

[2] For an extensive account of the various rate structures in 1934 for water in Chicago suburban areas, see Max R. White, Water Supply Organization in the Chicago Region (Chicago: University of Chicago Press, 1934).

[3] Interview with the Mayor of Burbank, August 10, 1971.

Street Light Districts

By an act of 1949 the Legislature authorized the establishment of street light districts in unincorporated territory. The district may contract for the installation, rental, or use of street lights within the district.

The district is governed by three trustees who serve for three-year staggered terms. The trustees are appointed by the chief executive officer of the Cook County Board, with the approval of the remaining board members. The trustees are empowered to levy a property tax; to issue bonds, upon approval of the voters at a referendum; to prescribe the work rules of district employees; and to make contracts for service. Statutory provisions do not exist for the dissolution of street light districts, but provisions exist for disconnection and annexation.[1]

The Norwood Park Street Light District is the only district of this type in Cook County. This tiny district has an area of only about one-third of a square mile. It is entirely residential in character and occupies an enclave on the northwest side of Chicago (Figure 18). The district obtains its lighting by contract with a private company.[2]

Summary

Fringe districts in Cook County have been formed largely since World War II in the newly urbanizing parts of the county. In addition to great population growth in outlying municipalities, much growth has taken place in former rural areas whose citizens desire urban services but do not want municipal government to provide them. Yet the existing general-purpose governments, counties and townships, have not responded in furnishing urban services. In resolving this dilemma, the formation of Special Districts has enabled the political system to respond to new demands, and at the same time forestalled the need for changing the structure of general-purpose government. Even where municipal incorporation or annexation has taken place, in many cases the district form of organization has been retained for providing certain urban services.

This chapter concludes the study of individual types of districts. The following chapter will deal with selected aspects of Special Districts in summary form.

---

[1] Illinois, Annotated Statutes, ch. 121, sec. 355-364.

[2] It proved impossible to contact officials of this district. Information was obtained in a conversation with the Supervisor of Norwood Park Township, August 16, 1972.

CHAPTER VII

IMPACT OF SPECIAL DISTRICTS ON THE

SPACE-POLITY OF COOK COUNTY

The preceding chapters have dealt with the characteristics of Special
Districts in three environmental settings structured around varying combinations of general-purpose governments. This chapter contains a more general
discussion of how Special Districts themselves function as governmental entities
and relate to the Cook County space-polity. Here we shall confine our attention
to five spatial aspects of this interrelationship: (1) the size of Special Districts;
(2) the distribution of resources; (3) horizontal integration; (4) centralization
and local control; and (5) an appraisal of Special Districts.

The Size of Special Districts

How large should a Special District be? According to Jackson and Bergman:

> The territorial jurisdiction of the government should cover and unite the
> activity field and the area of any specific problem to be solved. A political
> structure can function best when all interested parties and the entire area
> of the problem are included. [1]

Applying this normative guideline to the present study, we shall consider
"activity field" as the political area in which a service is actually being rendered. The term "problem area" will mean simply the area in which a service
is provided, regardless of the number or type of governments furnishing the
service.

If we examine the area in which any service is performed in the Chicago
area, then no Special District, nor any other governmental form, encompasses
sufficient territory so that the service is performed by a single government.
The degree of political integration necessary for this areal correspondence

---

[1] Jackson and Bergman, A Geography of Politics, p. 11.

between activity field and problem area is far from accomplished.

As we look deeper into the reasons why a district has a given size and why it does not fit the problem area, we find that it is a combination of many factors involving the type of service performed, the boundaries of other governmental forms, and the attitudes and goals of the district residents.

Only in the area-wide districts do we even have an approach to the ideal size. These districts deal with the functions of providing forest preserves, sewage disposal, tuberculosis control, and mosquito control. These functions are not desired by other governments because of their expense and areal unsuitability for small-scale operations. Moreover, the functions are remarkably uniform in quality, and there is little sentiment to provide differential levels of service. In summary, the area-wide districts are sufficiently large to afford economies of scale, and desires for local control are weakly expressed.

As large as the area-wide districts are, none of them extends beyond the boundary of Cook County.[1] There is little sentiment outside Cook County to combine functions with Cook County-based governments. Such a sentiment, however, does not preclude working arrangements with Cook County districts. For example, the Suburban Cook County Tuberculosis Sanitarium District has agreements to provide hospital care for tuberculosis patients from many neighboring counties.

When we turn our attention to the Special Districts other than area-wide districts, we become much more involved in vested interests and the attitudes of the people themselves. The residents of each district like to think themselves responsible for their own welfare, and concern for the welfare of other areas is little voiced. Stated another way, the efforts and concerns of the district officials and residents are directed to solving the problem as they view it. Unfortunately, each district sees only a small part of the problem. To an outsider, whose perceptual field is very likely different from that of residents of the district, the problem area and activity field may not coincide. However, to people within the district the activity field corresponds with their perception of the problem.

The location of district boundaries is seldom determined on the consideration of whether economies of scale can be obtained. Rather the boundaries are based on the judgment of the decision-makers regarding such factors as antici-

[1]The Regional Transit Authority (RTA) was created in the spring of 1974 for Cook County and five adjacent counties. This authority does not have the power of property taxation; therefore, it is not considered a Special District as defined in this study.

pated growth, areas from which support can be expected, and the existence of other nearby governmental forms. Such an emphasis on these latter considerations enhances local political control and leads to a large number of small-size districts rather than a single large-size district in which scale economies can be achieved. This failure to reconcile adequately a desire for effective local control and the appropriate scale of organization is a widespread problem in the United States and is well-stated in an article on government:

> The criteria of effective control of efficiency and of the inclusion of appropriate political interests, can be formulated on general theoretical grounds, but their application in any political system depends upon the particular institutions empowered to decide questions of scale. The conditions attending the organization of local governments in the United States usually require that these criteria be controlled by the decisions of the citizenry in the local community, i. e., subordinated to considerations of self-determination. [1]

When settlements were far apart, the district boundaries could be adjusted easily to the most appropriate service area. As time went on, however, the settlements became closer together, and the district boundaries abutted the boundaries of other districts. Compromises among interest groups became inevitable, and the activities of the district crystallized within static boundaries. When these boundaries are observed in today's landscape, they often appear incongruous because they represent a stage of occupance not evident today. However, if one examines the activity fields of the district interest groups, the boundaries become very real indeed.

A major source of support for many districts comes not only from the services they perform but also from the cultural role they play in the community. Many opportunities are afforded for social actors to become involved in community activities. Interaction among district residents provides a sense of belonging and security. This cultural role is best served by homogeneity in socioeconomic status and values, as well as an activity field in which interpersonal contact can be maintained. If a district becomes too large, it loses its homogeneity and becomes formalized in its organization; both of these factors may lead to a loss of local control and the support of the district residents.

Considerable variations in size exist among Special Districts, even when they are of the same type. The surprising discovery was that so many officials felt that their district was about the right size. This acceptance of the district

---

[1] Vincent Ostrom, Charles M. Tiebout, and Robert Warren, "The Organization of Government in Metropolitan Areas: A Theoretical Inquiry," American Political Science Review, LV (December, 1961), 837.

size suggests that the particular chain of historical circumstances that led to the formation of the district, as well as the legal difficulties in changing the present boundaries, have combined to condition the attitude of the officials as well as the district residents to accept the district as it exists. This interaction between the chain of historical circumstances and the attitudes of the people has solidified the boundaries of the district and helps to explain why the boundaries have remained static for such a long period of time.

The Distribution of Resources

If one considers all of Cook County, the resources are enormous. Within the county, however, the resources and needs are very unevenly distributed, and the present political structure tends to preserve or ignore the disparity. Here there is a conflict between the goals of equity and local control; and in the case of Special Districts, the results are mixed, although local control appears to dominate.

The area-wide districts come closest to the goals of equity and efficient operations. They have the most taxable property at their disposal, can achieve economies of scale, and can move resources easily from surplus areas to deficit areas. For example, the Forest Preserve District has received the larger part of its tax resources from Chicago; yet the funds have been spent mostly in suburban areas where the forests are located. This movement of funds permits the citizens of Chicago to establish forests in areas where they are most suited and still benefit from them. The suburban areas also benefit, because without the Chicago funds, the forest preserves could not have been as extensive as they now are. Similarly, mosquito abatement districts spend much of their effort in swampy and wooded areas that yield little in revenue; yet the benefits from the control measures spill over into the revenue-producing areas. In summary, the area-wide districts make it possible to internalize the benefits and costs within the same political jurisdiction.

Once we drop below the scale of area-wide districts, it is much more difficult to match up resources and needs, achieve economies of scale, and avoid the spill-over of costs and benefits across district boundaries. The unequal distribution of resources results from differences in location, socioeconomic status, degree of development, amount of taxable property, and degree of obsolescence. Most of the districts receive the bulk of their income from taxation; only the collector sanitary districts receive the larger part of their income from user charges.[1] Since most district income comes from taxation, the presence or absence of taxable property within the district boundaries great-ly affects the capability of the district to perform its functions.

In practice, each district adapts its functions to its size and available resources. If a district is too small for scale economies and still limited in its resources, it simply lowers the quality of services. For example, the small-est districts have no separate offices and no full-time personnel, and services are minimal. At the other extreme, districts that contain much taxable prop-erty or are large enough to achieve economies of scale may have elaborate facil-ities and a large staff of technically trained personnel. To some extent, the lack of tax resources is supplemented by a schedule of fees, which vary accord-ing to the socioeconomic status of the district residents. In summary, the pres-ent system of district organization permits great inequalities to exist in facil-ities and the rendering of services. Presumably, the lack of uniformity and lack of equity meet with the approval of the district residents, the Legislature, and the courts.

Still another factor affecting the distribution of resources is the spill-over of benefits and costs across district boundaries. For example, the Chi-cago Park District maintains facilities that are of use to areas far beyond the boundaries of the district. Indirectly, the presence of these facilities enables suburban park districts to have fewer facilities because suburban residents can travel to the Chicago Park District for facilities that are not available in sub-urban districts. Another type of spill-over is indicated by the presence of large municipal fire departments that provide additional protection to small fire pro-tection districts. If a conflagration becomes too great and exceeds the capacity of the district equipment and personnel, then the municipal fire department can be called on for assistance.

Horizontal Integration

The structure of local government in Cook County is incredibly complex. Thirty-eight townships, 128 municipalities, 150 school districts, and 196 Spe-cial Districts fragment the county into a multi-nucleated system of decision-making. The territories of these governments overlap one another in various ways, resulting in thousands of small pieces of territory, each piece with its

---

[1]Generalization obtained from a perusal of district expenditures and in-come contained in U.S. Bureau of the Census, Census of Governments: 1962, V, 505-6. Financial data on mass transit districts are not available.

own tax rate, debt structure, and combination of officials.

Such a lack of an orderly arrangement of local governments in Cook County was built into the structure from early times. Through decades of homeostatic adjustment, the political system of Cook County has been characterized by increasing the number of decision-making units. The space-polity has been continually modified as one government after another was superimposed over existing governments. The existing governments were not necessarily abolished; but as each new government was added, there were varying degrees of readjustment of functions and governmental areas.

The description of interrelationships among these governments can be facilitated by introducing the concept of horizontal ordering which deals with the arrangement of governments at the same level. In a legal sense, counties, municipalities, school districts, and Special Districts are coordinate governments since they all derive their powers from the State of Illinois. For the most part, no government is subordinate to the other;[1] only the townships are subject to the county. The convenient hierarchical ordering (vertical ordering) of governments, in which lines of authority from one level to another are clearly defined, simply does not exist in Cook County. Since most Special Districts in Cook County are independent of other governments, nearly all integration involving Special Districts is of the horizontal type.

When horizontal integration is examined spatially, there often is a problem of overlapping territory. Neither a Special District nor a general-purpose government may overlap another government of the same type. However, a Special District or a general-purpose government may overlap a government of another type.[2] For example, a municipality, township, park district, and fire protection district may all occupy the same territory.

For convenience in discussion, the analysis of horizontal integration can be divided into two broad categories: functional and areal. First, we can view a particular function and determine the degree of coordination among the differ-

---

[1]The Forest Preserve District, Stickney Public Health District, and Berwyn Public Health District are considered as dependent governments by the Census Bureau. Even though they are dependent governments, these districts are distinct legal entities and have relationships with other governments, just as independent districts do. Consequently, their inclusion in the discussion of horizontal integration seems amply justified.

[2]For a general discussion of this topic, see Jackson and Bergman, A Geography of Politics, pp. 8-10. According to Illinois law, sanitary districts of the 1936 type may overlap sanitary districts formed under the laws of 1917 and 1889.

ent areas receiving the same service regardless of the governmental structure furnishing the service. Second, we can view a unit of area and determine the coordination among the different functions being rendered in the same place.

Functional Horizontal Integration. --Each function must be examined individually as to the degree of integration that exists among the governmental forms providing the service. The function may be performed by a single large government or by a number of small governments. Moreover, these functions may be performed by general-purpose governments, Special Districts, or a combination of both.

In Cook County the districts that bring about functional cooperation among separate general-purpose governments are one of the important factors of functional horizontal integration.[1] Through the mechanism of the district device, the general-purpose governments can be integrated for some functions but not others. Areas under different municipal governments can be combined with each other or with unincorporated areas under township government. Yet the units of general-purpose government can remain independent through the control of other services. Moreover, the political organizations, which are organized around areas of general-purpose government can remain intact.

In order to arrive at a quantitative assessment of this functional cooperation, maps of Cook County were examined to determine how many municipalities were included within a single district.[2] In this manner, it can be determined how integration varies with the type of function. As is shown in Table 14, the greatest degree of functional cooperation is in providing forest preserves, sewage disposal, tuberculosis care, and mosquito control. Not surprisingly, these functions are the ones provided by the area-wide districts. For these functions, which are not desired by other governments, we could say that Cook

---

[1] Jacob and Teune list ten factors as leading to integration. These are: (1) proximity; (2) homogeneity; (3) transactions; (4) mutual knowledge; (5) functional interest; (6) communal character; (7) political structure; (8) sovereignty; (9) governmental effectiveness; and (10) integrative experiences. Special Districts manifest several of these factors, but here we are concerned with how people are brought together because they share a particular service rendered by a Special District. Source: Jacob and Toscano (eds.), The Integration of Political Communities, pp. 16-44.

[2] To be counted as an integrating district, the district must contain more than half the territory of two or more municipalities. For those districts straddling the Cook County boundary, all of the district is examined, even if the municipalities within the district lie outside Cook County.

TABLE 14

SPECIAL DISTRICTS COVERING MORE
THAN ONE MUNICIPALITY

Type	Number	No. of Municipalities Within Districts[a]
Forest Preserve	1	119
Tuberculosis Care	1	118
Sewage Disposal (1889 type)	1	107
Mosquito Abatement	4	90
Mass Transit	4	40
Park	10	20
Fire Protection	5	11
Library	3	7
Public Health	1	4
Sewage Disposal (1917 type)	1	3

[a]More than one-half of municipality must
lie within district to be counted.

Source: Compiled from maps of Illinois Department of Local Government Affairs.

County is fairly well integrated. Working arrangements with other districts
and general-purpose governments have been worked out over time, and the present arrangement seems to meet with the approval of most citizens. On the
other hand, the emphasis on the service-by-service solution to problems has
tended to split the space-polity along functional lines, and integration among
these functions is relatively small. Concerning these area-wide districts,
Pock remarks:

> It can hardly be denied that metropolitan districts, by their ability to strad-
> dle boundary lines, have been successful in bringing about a horizontal inte-
> gration of certain urban functions with which they have been entrusted.[1]

The district device also permits cooperation among areas of general-
purpose government for other functions but to a much lesser extent than those
provided by the area-wide districts. The largest of these are the relatively
new mass transit districts which permit municipalities to improve transportation facilities through cooperative effort. In a few cases, two or more small

[1]Pock, Independent Special Districts, p. 79.

municipalities lie entirely within a single park district, library district, public health district, or fire protection district.

Except for the area-wide districts and a few others, locally based districts are the rule. There is a host of park, fire protection, and collector sanitary districts with very limited jurisdictions. The degree of horizontal integration among these districts is small, and any coordinated program among them is virtually impossible.

Areal Horizontal Integration. --The integration among different functions varies from place to place throughout the county. In each small area of Cook County an ad hoc arrangement is made among all the governments providing services. A few services are provided by the previously mentioned area-wide districts, some are provided by general-purpose governments, and some by local Special Districts. At times the same service is provided by a Special District and a municipality in the same area. An over-all evaluation for this type of integration would be poor. There is little sentiment for a coordinated policy; and in each small area, the local citizens apparently believe they are capable of making the best arrangements to satisfy the needs for services.

The most integration exists within Chicago which has a single municipal government with no township organization. District government in Chicago is fairly well consolidated, consisting of a single park district for the whole city and a mass transit district for the downtown area; of course, the city also lies within the Forest Preserve District and the MSDGC. A mosquito abatement district covers the southernmost part of the city.

For the rest of Cook County outside Chicago extensive fragmentation exists among both general-purpose governments and Special Districts. A tangled mass of park districts, fire protection districts, library districts, and others are superimposed over an expanse of zig-zag municipal boundaries.

With such extensive fragmentation, one might wonder how Cook County operates at all. In practice, the situation is much better than the territorial arrangement might suggest. Through time the districts have developed a series of understandings, agreements, and working arrangements with other governments and with each other. Local officials believe that this horizontal integration leads to a superior type of government. It prevents the accumulation of power in a few centers and insures that programs are really locally applicable.

Although this intergovernmental cooperation has been going on for a long time, it is expected to increase within the next few years. As noted in a recent report:

Though intergovernmental agreements were legal before 1970, their potential usefulness was greatly augmented by the broadened powers of the new Illinois Constitution. Article VII, Section 10, of the Constitution provides vast new authority for units of local government to team up through cooperative agreements. Constitutional Convention delegates clearly intended to reduce local government's excessive dependency on the state by strengthening its hand in problem solving.

Under the new Constitution and the Intergovernmental Cooperation Act, units of local government--cities, villages, incorporated towns, counties, townships, special districts and other governmental bodies so designated by law--may enter into intergovernmental agreements . . . with each other, with the state, the United States, and even with individual and private enterprises.[1]

Great skill in management is required in these complex arrangements. Harlan Cleveland writes of our "horizontal society," and points out the difficulties in coordination. As he says:

At every level of government the increasing complexity of the subject matter, plus the increasing sensitivity to the interdependence of domains previously accepted as fairly separable, multiples the number of executives whose special knowledge is essential or whose oxen are gored. In every community, and notably in metropolitan areas, a new pattern of leadership now spreads the power to affect the community's destiny, breaking the leadership monopolies traditionally held by businessmen, business lawyers, and early-arriving ethnic groups. In the new competition for influence, the ticket of admission for the leaders of any aspiring group is now skill in organization, and a working knowledge of interorganizational complexity. For every new decision--a new hospital, a downtown plaza, a poverty program, a community college, a metropolitan water plan, or whatever--involves the creative management of multiple authorities, "private" as well as public.[2]

Centralization and Local Control

Cook County long has been the battleground between advocates of centralization and advocates of local control. Neither of these viewpoints has been completely dominant, and virtues accompanying one or the other have occupied both reformers and politicians for decades. As has been shown previously, Special Districts reflect compromises between both viewpoints, varying with the type of service provided and with the particular combination of general-purpose governments the district serves. In this section, we shall examine some of the advantages and disadvantages contained in both points of view.

---

[1] Northeastern Illinois Planning Commission, An Introduction to Inter-Governmental Agreements (Chicago: Northeastern Illinois Planning Commission, 1974), p. 1.

[2] Harlan Cleveland, The Future Executive (New York: Harper & Row, 1972), p. 32.

The Case for Centralization. --Hirsch mentions three major advantages that have been claimed for consolidation of local governments: (1) savings due to economies of scale, (2) improved conditions for coordinated planning and orderly growth, and (3) equity in financing governmental services.[1]

The concept of economy of scale is a useful device to show how costs per unit of output vary with the size of the administrative area producing the output. Presumably, for each service there is an optimal size of area in which the lowest cost per unit is achieved, and making the administrative area larger or smaller than this optimal size increases the cost per unit.[2] Unfortunately, detailed data are not available that would show how the cost per unit varies with the size of district in Cook County. That the alleged benefits from scale economies is a muddled issue is vividly pointed out by Walsh:

> The degree to which the operation of large-scale services is in fact more economical than the operation of smaller ones has not been proven and certainly varies with the type of service. . . . given an urban complex of a certain size and shape, there is no evidence to show that per capita costs of providing services would be greater or less if it were governed by one large unit or by several smaller ones.[3]

Conversations with district officials reveal that many districts in Cook County apparently are sufficiently large to achieve economies of scale. The consolidation of these districts might achieve greater equity or coordination but would not provide lowered costs per unit of output. Only for the smallest districts could it be suggested that they do not achieve economies of scale. Just what the optimal size for each type of district would be is a subject worthy of further investigation.

Another claimed virtue for consolidation relates to coordinated and orderly planning for growth. Through consolidation, spill-over benefits and costs from smaller units would still be within the larger unit and could be taken into consideration in planning. The planning for services provided by Special Districts in Cook County has been extremely limited. Only the Forest Preserve District grew up as direct result of planning. Although not primarily oriented towards Special Districts, some planning has been done for Cook Coun-

---

[1] Werner Z. Hirsch, "The Urban Challenge to Governments, " in America's Cities, Michigan Business Papers No. 54, ed. by Wayland D. Gardner (Ann Arbor: University of Michigan, Bureau of Business Research, 1970), p. 48.

[2] Walter Isard, Methods of Regional Analysis: An Introduction to Regional Science (Cambridge, Mass.: M. I. T. Press, 1960), pp. 527-28.

[3] Annmarie Hauck Walsh, The Urban Challenge to Government: An International Comparison of Thirteen Cities (New York: Frederick A. Praeger, 1969), pp. 53-54.

ty and adjacent counties for sewage disposal, waste water, and open space. For the most part, however, districts for functions such as parks, fire protection, sewage collection, libraries, and mass transit have grown up as needs were perceived by locally based interest groups.

The consideration of equity (or lack of it) has been alluded to in previous sections of this paper. The great differences in socioeconomic status and taxable property from one micro-area of Cook County to another are paralleled by great differences in the quality of services provided by Special Districts that are not area-wide in scope. Services for parks and fire protection, for example, vary from almost none to excellent. The fragmentation of park and fire protection districts is such that it makes any uniform standard practically impossible of achievement.

The Case for Local Control. --In addition to the advantages claimed for centralization, advantages also can be claimed for local control. Under local control: (1) government can be made more responsive; (2) competition can be introduced into the public sector; and (3) corruption in government can be more easily controlled.

Local control increases the responsiveness of government by providing easy access to local officials who have more knowledge of the unique situations involved in any area. It is possible for problems to be handled with a minimum of bureaucratic delay. Many district officials maintain that autonomy actually makes them more efficient, for they are not subject to arbitrary regulations or budgetary restrictions of general-purpose government. In an appraisal of a large metropolitan government called "gargantua," it is noted:

> However, gargantua with its single dominant center of decision-making, is apt to become a victim of the complexity of its own hierarchical or bureaucratic structure. Its complex channels of communication may make its administration unresponsive to many of the more localized public interests in the community. The costs of maintaining control in gargantua's public service may be so great that its production of public goods becomes grossly inefficient. . . . Bureaucratic unresponsiveness in gargantua may produce frustration and cynicism on the part of the local citizen who finds no point of access for remedying local problems of a public character. . . . Large-scale, metropolitan-wide organization is unquestionably appropriate for a limited number of public services, but it is not the most appropriate scale of organization for the provision of all public services required in a metropolis.[1]

---

[1] Ostrom, Tiebout, and Warren, "Organization of Government in Metropolitan Areas," pp. 837-38.

Another claimed virtue for local control would be the opportunity for small jurisdictions to make choices among several governments offering the same services. According to this viewpoint, the costs of duplication would be more than offset by the benefits derived from introducing competition into the public sector. As noted in an article:

Patterns of competition among producers of public services in a metropolitan area, just as among firms in the market, may produce substantial benefits by inducing self-regulating tendencies with pressure for the more efficient solution in the operation of the whole system. Variety in service levels among various independent local government agencies within a larger metropolitan community may give rise to a quasi-market choice for local residents in permitting them to select the particular community in the metropolitan area that most closely approximates the public service levels they desire. Public service agencies then may be forced to compete over the service levels offered in relation to the taxes charged. [1]

Many governments in Cook County have contracts with other governments for certain services. For example, some fire protection districts obtain their protection from municipal fire departments. Also some collector sanitary districts obtain water from municipal water systems. It is difficult to say to what extent the charges that districts pay are influenced by competition, although it is safe to say that the charges would not be unreasonable.

The presence of corruption, or the appearance of it, in certain large centralized governments is probably a factor explaining why there is little sentiment for centralizing authority in a few centers. As soon as governments are consolidated, it gives authority to the top people in the hierarchy. Judging from impressions obtained in interviews, it evidently is felt that consolidation of power into a larger unit of government merely enhances the prestige and perquisites of those controlling the larger unit of government to the detriment of individual units. Consolidated governments as exemplified by the Chicago and Cook County Governments, have proved to be unattractive models for the people of Cook County in recent decades. A political system with many centers of decision-making is a protection against the abuse of power. [2]

Summary.--The foregoing arguments illustrate that the size and level of government which can best produce a particular public good or service is a com-

---

[1] Ibid.

[2] For a critical commentary on the shortcomings of centralized government as it now functions, see: Lee Ahlswede, "Examining Metro Government Today," Illinois County & Township Official, XXXIV (May, 1974), 24-26.

plex question. The gains and losses of decentralized authority need to be weighed and traded off in the light of the following criteria: efficiency through scale economies, orderly planning and allocation of resources, equity in financing and distributing urban governmental services, responsiveness of government to public demands, opportunities for choice, and the maintenance of honesty and integrity in government. Unfortunately, no method has been developed which interrelates all these factors on a common scale.

In Cook County the particular combination of the above-mentioned criteria differs from district to district. Obviously, all of the criteria cannot be maximized simultaneously; and, according to Illinois law, the trades-off among the various criteria are determined primarily by the citizens of the local district.

An Appraisal of Special Districts

Theoretically, it would be possible to adapt general-purpose governments to such an extent that Special Districts would not be necessary. Such adaptation, however, would require extensive reforms in public attitudes and societal institutions, at best a slow and difficult process. Thus, in a realistic sense, Special Districts are effective instruments for circumventing many dysfunctional aspects of society such as conflicting interests, restrictive laws, inappropriate boundaries of existing governments, and unresolved questions of equity and local control. Vital services are performed, the resolution of contentious and difficult questions is avoided, traditional governments are not threatened, and local control over selected services is maintained.

Both advantages and disadvantages can be ascribed to Special Districts in Cook County.[1] The evidence suggests that the advantages outweigh the disadvantages, for the districts enjoy substantial support and evidently are meeting the needs and fulfilling the goals of Cook County citizens. Judging from interviews, there is far from a universal commitment to implement egalitarian goals, plan for orderly growth, or achieve the most efficient size of organization for providing services. In addition to the provision of services at a satisfactory level, people desire a decentralization of power, accompanied by a government that is accessible, responsive to demands, free from the discipline of

---

[1]For a discussion of the conflicting views towards Special Districts, see a collection of papers in "Public Administration Forum: The Special Districts as Instrumentality [sic] of State and Local Government," Midwest Review of Public Administration, I (August, 1967), 119-33.

party organizations, and one in which they can participate.

Disadvantages such as a loss of scale economies, difficulties in coordination, and a lack of planning have been noted by numerous writers; but to infer that these dysfunctional elements would disappear if Special Districts were eliminated is patently untrue. The fact that these elements also exist in general-purpose governments suggests that Special Districts are the manifestations rather than the cause of some arbitrarily defined social malaise.

Even if it were desirable to eliminate Special Districts in Cook County, the political questions involved in dissolving or consolidating them are formidable ones. Little sentiment has been noted for eliminating districts and assigning their functions to other governments. Just as interest groups were instrumental in forming districts, interest groups would have to be active in their dissolution. This seems unlikely to happen as the interest groups also form support for the districts. Only collector sanitary districts and fire protection districts, under certain conditions, may be dissolved on the initiative of municipal authorities.

The new Illinois Constitution of 1970 removes many of the legal barriers that have led to the formation of numerous Special Districts in Cook County. Debt restrictions have been eased considerably, and proscriptions against differential taxing areas have been removed. But since these legal barriers comprise only part of the constraints coming from environmental sub-systems, new districts probably will continue to be formed. Constraints relating to the authorization of powers for local governmental units, the areal convenience of the district device, the resistance to municipal government, and the desire of citizens to isolate themselves from general-purpose governments probably will continue to be influential in the formation of new districts.

APPENDIX I

NOTES ON THE FIGURES

Figure
1. The diagram has been adapted with modifications from Easton, A Systems Analysis of Political Life, pp. 30, 37, and 381; and also from Walsh, Urban Challenge to Government, p. 224.

2. The townships were taken from untitled tax maps of Cook County, prepared by the Illinois Department of Local Government Affairs, 1971-1973.

3. The municipalities were taken from "Map of Cook County, Illinois, 1973, Showing Highways and Forest Preserves," issued by the Board of Commissioners of Cook County.

4. Data on the growth of the MSDGC were assembled from a map, "Annexations and Municipalities," prepared by the MSDGC, 1973.

5. Data on waterways of the MSDGC were assembled from "Map of the Metropolitan Sanitary District of Greater Chicago, Showing Sewage Treatment Works, Pumping Stations, Water Reclamation Plants, Retention Reservoirs, Channels, and Sewers," prepared by the MSDGC, 1971.

6. The locations of the sanitarium, clinics, and cooperating hospitals were taken from the "1969 Annual Report," prepared by the SCCTSD.

7. The mosquito abatement districts were taken from untitled tax maps of Cook County, prepared by the Illinois Department of Local Government Affairs, 1971-1973.

8. The park districts were taken from untitled tax maps of Cook County, prepared by the Illinois Department of Local Government Affairs, 1971-1973.

9. The area of the original park districts (1869 area) was taken from descriptions in Private Laws of the State of Illinois (1869), I, 344, 358, 376, and 377. The areal additions from 1870 to 1917 were derived from a map in Chicago Bureau of Public Efficiency, Nineteen Local Governments in Chicago, p. 27. The areal additions from 1918 to 1939 were derived from the Illinois Tax Commission, Atlas of Taxing Units, pp. 47-52. A map showing the park districts in 1962 could not be found; consequently, the areal additions between 1940 and 1962 were estimated. New park districts, formed between 1940 and 1962, were noted in the Census of Governments: 1962, V, 507-8; however, the area of these new

districts was assumed to be the same as it was in 1973. It also was assumed that the changes in area between 1940 and 1973 for old districts (those formed prior to 1940) took place between 1940 and 1962. The areal additions between 1963 and 1973 were assumed to be made up of new districts (formed after 1962). Exceptions were made for Chicago where the boundary of the Chicago Park District changed with the city boundary. In Chicago, therefore, the park district boundary could be determined by noting the city boundary changes on "Map of Chicago: Showing Growth of the City by Annexations and Accretions, " prepared by the Department of Public Works.

Figure
10.   The location of parks and the park district boundary was taken from a map in Mount Prospect Park District, "Fun Talk, " Vol. II (July-August, 1971). Municipal boundaries were taken from untitled tax maps of Cook County, prepared by the Illinois Department of Local Government Affairs, 1971-1973.

11.   The three original park districts were compiled from information in Private Laws of the State of Illinois (1869), I, 344, 358, 376, and 377.

12.   The boundaries of public health, outlying sanitary, and river conservancy districts were taken from untitled tax maps of Cook County, prepared by the Illinois Department of Local Government Affairs, 1971-1973.

13.   The library districts were taken from untitled tax maps of Cook County, prepared by the Illinois Department of Local Government Affairs, 1971-1973.

14.   The identification of mass transit districts was taken from the legends of untitled tax maps of Cook County, prepared by the Illinois Department of Local Government Affairs, 1971-1973. The base map for municipalities comprising the districts was prepared from "Map of Cook County, Illinois, 1973, Showing Highways and Forest Preserves, " issued by the Board of Commissioners of Cook County.

15.   The map of the Chicago South Suburban Mass Transit District was prepared from the same source noted for Figure 3.

16.   The fire protection districts were taken from untitled tax maps of Cook County, prepared by the Illinois Department of Local Government Affairs, 1971-1973. The following districts, although shown on the tax maps, are dissolved and are not shown on Figure 16: Dixie Gardens, Grandview Park, and Rolling Meadows.

17.   Data for compilation of map of Barrington Countryside Fire Protection District was obtained from untitled tax maps of Cook County, Kane County, Lake County, and McHenry County, prepared by the Illinois Department of Local Government Affairs, 1971-1973.

18.   The map of collector sanitary and street light districts was taken from untitled tax maps of Cook County, prepared by the Illinois Department of Local Government Affairs, 1971-1973. The following collector sanitary districts, although shown on the tax maps, are dissolved and are not

shown on Figure 18: Grandview Park, Manor Heights, Norwood Park, and Ridgeland Park. The Norwood Park Street Light District is shown on the tax maps as occupying both enclaves in northwestern Chicago. Actually, the district occupies only the northernmost enclave.

Figure
19. The map of South Stickney Sanitary District was prepared from untitled tax maps of Cook County, prepared by the Illinois Department of Local Government Affairs, 1971-1973. The locations of the water system and sewer system connections were obtained from the Superintendent of South Stickney Sanitary District, August 2, 1970.

APPENDIX II

NOTES ON THE TABLES

Table

6. [a]Alice Greenacre, A Handbook for the Woman Voters of Illinois (Chicago: Chicago School of Civics and Philanthropy, 1913), p. 53.

[b]Illinois Tax Commission, Atlas of Taxing Units, p. 21.

[c]Chicago Bureau of Public Efficiency, Unification of Local Governments in Chicago, pp. 20-23.

[d]Compiled from Jensen, "Financial Statistics in Cook County," pp. 4-31.

[e]Compiled from maps in Illinois Tax Commission, Atlas of Taxing Units, pp. 47-52.

[f]U.S. Bureau of the Census, Census of Governments: 1962, V, 507-8.

[g]Compiled from untitled tax maps of Cook County, prepared by the Illinois Department of Local Government Affairs, 1971-1973.

7. [a]The area of park districts was compiled from descriptions in Private Laws of the State of Illinois (1869), I, 344, 358, 376, 377.

[b]The area of park districts was compiled from sources in footnote a, modified by boundary changes noted on "Map of Chicago: Showing Growth of the City by Annexations and Accretions," prepared by the Department of Public Works. The area of the Chicago Sanitary District was compiled from a map "Annexations and Municipalities," prepared by the MSDGC.

[c]The area of the twelve park districts in Chicago was measured from a map in Chicago Bureau of Public Efficiency, Nineteen Local Governments in Chicago, p. 27. For the three small park districts in Chicago formed between 1915 and 1917, the average area for small park districts of 2.7 square miles was used. For the eleven suburban park districts, the 1939 area was used. The area of the Chicago Sanitary District was compiled from the map cited in footnote b.

[d]The area was measured from maps in Illinois Tax Commission, Atlas of Taxing Units, pp. 47-52.

Table
7.   [e]The area was measured from untitled tax maps of Cook County, prepared by the Illinois Department of Local Government Affairs, 1971–1973.

[f]The area is less than 0.5 square mile.

11.   [a]The number of park districts was compiled from the same sources as listed in footnotes to Table 6. The area of park districts was compiled from the same sources as listed in footnotes to Table 7.

12.   [a]Except where otherwise noted, the assessed valuation and population data for individual park districts were taken from Northeastern Illinois Planning Commission, Suburban Factbook, pp. 116-18.

[b]Barrington Countryside Park District is nearly coterminous with the municipality of Barrington Hills. The population of the municipality was taken from U.S. Bureau of the Census, Census of Population: 1970. Phoenix Park District is coterminous with the municipality of Phoenix. The population of the municipality was taken from the previously cited source.

[c]Hawthorne Park District and Clyde Park District cover all of the municipality of Cicero. The population of Hawthorne Park District was estimated by deducting the population of Clyde Park District from the population of Cicero.

[d]River Forest Park District is coterminous with Cook County Elementary School District No. 90. The assessed valuation for the school district is taken from Northeastern Illinois Planning Commission, Suburban Factbook, p. 105.

[e]The population data for Golf-Maine Park District was taken from "1971 Illinois Association of Park Districts Directory," Illinois Parks and Recreation, II (September-October, 1971).

SELECTED BIBLIOGRAPHY

Books--General

Advisory Commission on Intergovernmental Relations. Performance of Urban Functions: Local and Areawide. Washington: U.S. Government Printing Office, 1963.

_____. Substate Regionalism and the Federal System. Vol. I: Regional Decision Making: New Strategies for Substate Districts. Washington: U.S. Government Printing Office, 1973.

Alderfer, Harold F. American Local Government and Administration. New York: Macmillan Co., 1956.

Anderson, William. The Units of Government in the United States. Rev. ed. Chicago: Public Administration Service, 1949.

Banfield, Edward C., ed. Urban Government: A Reader in Administration and Politics. Rev. ed. New York: Free Press, 1969.

Bollens, John C., ed. Exploring the Metropolitan Community. Berkeley: University of California Press, 1964.

Bollens, John C.; Bayes, John R.; and Utter, Kathryn L. American County Government: With an Annotated Bibliography. Beverly Hills, Cal.: Sage Publications, 1969.

Bollens, John C. and Schmandt, Henry J. The Metropolis: Its People, Politics, and Economic Life. 2nd ed. New York: Harper & Row, 1970.

Bonjean, Charles M.; Clark, Terry N.; and Lineberry, Robert L.; eds. Community Politics: A Behavioral Approach. New York: Free Press, 1971.

Bromage, Arthur W. Political Representation in Metropolitan Agencies. Michigan Governmental Studies No. 42. Ann Arbor: University of Michigan, Institute of Public Administration, 1962.

Brown, Robert Harold. Political Areal-Functional Organization: With Special Reference to St. Cloud, Minnesota. Department of Geography Research Paper No. 51. Chicago: University of Chicago, 1957.

Buckley, Walter, ed. Modern Systems Research for the Behavioral Scientist: A Sourcebook. Chicago: Aldine Publishing Co., 1968.

Caiden, Gerald F. Administrative Reform. Chicago: Aldine Publishing Co., 1969.

164

Cleveland, Harlan. The Future Executive. New York: Harper & Row, 1972.

Cox, Kevin R. Conflict, Power, and Politics in the City: A Geographic View. New York: McGraw-Hill Book Co., 1973.

Duncombe, Herbert Sydney. County Government in America. Washington: National Association of Counties Research Foundation, 1966.

Easton, David. A Framework for Political Analysis. Englewood Cliffs, N.J.: Prentice-Hall, 1965.

_____. A Systems Analysis of Political Life. New York: John Wiley & Sons, 1965.

Elazar, Daniel J. Cities of the Prairie: The Metropolitan Frontier and American Politics. New York: Basic Books, 1970.

Fesler, James W. Area and Administration. University, Ala.: University of Alabama Press, 1949.

Gottmann, Jean and Harper, Robert A., eds. Metropolis on the Move: Geographers Look at Urban Sprawl. New York: John Wiley & Sons, 1967.

Greer, Scott. The Emerging City: Myth and Reality. New York: Free Press, 1962.

_____. Governing the Metropolis. New York: John Wiley & Sons, 1962.

Highsaw, Robert B. and Dyer, John A. Conflict and Change in Local Government: Patterns of Co-operation. University, Ala.: University of Alabama Press, 1965.

Hirsch, Werner Z. "The Urban Challenge to Governments." America's Cities. Michigan Business Papers No. 54. Edited by Wayland D. Gardner. Ann Arbor: University of Michigan, Bureau of Business Research, 1970.

Isard, Walter. Methods of Regional Analysis: An Introduction to Regional Science. Cambridge, Mass.: M.I.T. Press, 1960.

Isard, Walter and Coughlin, Robert. Municipal Costs and Revenues Resulting from Community Growth. Wellesley, Mass.: Chandler-Davis Co., 1957.

Jackson, W. A. Douglas and Bergman, Edward F. A Geography of Politics. Dubuque, Ia.: William C. Brown Co., 1973.

Jackson, W. A. Douglas and Samuels, Marwyn S., eds. Politics and Geographic Relationships. 2nd ed. Englewood Cliffs, N.J.: Prentice-Hall, 1971.

Jacob, Philip E. and Toscano, James W., eds. The Integration of Political Communities. Philadelphia and New York: J. B. Lippincott Co., 1964.

Kaplan, Harold. Urban Political Systems: A Functional Analysis of Metro Toronto. New York: Columbia University Press, 1967.

Kasperson, Roger E. and Minghi, Julian V., eds. The Structure of Political Geography. Chicago: Aldine Publishing Co., 1969.

Kotler, Milton. Neighborhood Government: The Local Foundations of Political Life. Indianapolis, Ind.: Bobbs-Merrill Co., 1969.

Lineberry, Robert L. and Sharkansky, Ira. Urban Politics and Public Policy. New York: Harper & Row, 1970.

Maass, Arthur, ed. Area and Power: A Theory of Local Government. Glencoe, Ill.: Free Press, 1959.

Maddox, Russell W., Jr., ed. Issues in State and Local Government: Selected Readings. Princeton, N.J.: D. Van Nostrand Co., 1965.

Massam, Bryan H. The Spatial Structure of Administrative Systems. Commission on College Geography Research Paper No. 12. Washington: Association of American Geographers, 1972.

Morgan, David R. and Kirkpatrick, Samuel A., eds. Urban Political Analysis: A Systems Approach. New York: Free Press, 1972.

Murphy, Raymond E. The American City: An Urban Geography. New York: McGraw-Hill Book Co., 1966.

Murphy, Thomas P. Metropolitics and the Urban County. Washington: Washington National Press, 1970.

National Municipal League. The Government of Metropolitan Areas in the United States. New York: National Municipal League, 1930.

Olsen, Marvin E. The Process of Social Organization. New York: Holt, Rinehart and Winston, 1968.

Platt, Robert S. A Geographical Study of the Dutch-German Border. Münster-Westfalen, W. Germany: Geographische Kommission, 1958.

Pounds, Norman J. G. Political Geography. 2nd ed. New York: McGraw-Hill Book Co., 1972.

Prescott, J. R. V. The Geography of Frontiers and Boundaries. Chicago: Aldine Publishing Co., 1965.

_____. The Geography of State Policies. Chicago: Aldine Publishing Co., 1968.

Rapoport, Anatol. "General Systems Theory." International Encyclopedia of the Social Sciences. 1968. Vol. XV.

Riker, William H. The Study of Local Politics. New York: Random House, 1959.

Ross, Russell M. and Millsap, Kenneth F. State and Local Government and Administration. New York: Ronald Press Co., 1966.

Scott, Mel. American City Planning since 1890. Berkeley: University of California Press, 1969.

Shank, Alan, ed. Political Power and the Urban Crisis. Boston: Holbrook Press, 1969.

Simon, Herbert A. Administrative Behavior: A Study of Decision-Making Processes in Administrative Organization. Glencoe, Ill.: Free Press, 1957.

Smallwood, Frank. Greater London: The Politics of Metropolitan Reform. Indianapolis, Ind.: Bobbs-Merrill Co., 1965.

Sofen, Edward. The Miami Metropolitan Experiment. Anchor Books. Garden City, N.Y.: Doubleday & Co., 1966.

Soja, Edward W. The Political Organization of Space. Commission on College Geography Resource Paper No. 8. Washington: Association of American Geographers, 1971.

Stagner, Ross. "Homeostasis." International Encyclopedia of the Social Sciences. 1968. Vol. VI.

Wakstein, Allen M., ed. The Urbanization of America: An Historical Anthology. Boston: Houghton Mifflin Co., 1970.

Walsh, Annmarie Hauck. The Urban Challenge to Government: An International Comparison of Thirteen Cities. New York: Frederick A. Praeger, 1969.

Whitney, Joseph B. R. China: Area, Administration, and Nation Building. Department of Geography Research Paper No. 123. Chicago: University of Chicago, Department of Geography, 1969.

Willbern, York. The Withering Away of the City. Midland Books. Bloomington, Ind.: Indiana University Press, 1966.

Wood, Robert C. 1400 Governments: The Political Economy of the New York Metropolitan Region. Anchor Books. Garden City, N.Y.: Doubleday & Co., 1964.

_____. Suburbia: Its People and Their Politics. Boston: Houghton Mifflin Co., 1958.

Books--Special Districts

Advisory Commission on Intergovernmental Relations. The Problem of Special Districts in American Government. Washington: U.S. Government Printing Office, 1964.

Asseff, Emmett. Special Districts in Louisiana. Baton Rouge: Louisiana State University, Bureau of Government Research, 1951.

Bain, Henry. The Development District: A Governmental Institution for the Better Organization of the Urban Development Process in the Bi-County Region. Report for the Maryland-National Capital Park and Planning Commission. Riverdale, Md.: Washington Center for Metropolitan Studies, 1968.

Bird, Frederick L. Local Special Districts and Authorities in Rhode Island.
Research Series No. 4. Kingston, R.I.: University of Rhode Island,
Bureau of Government Research, 1962.

Bollens, John C. Special District Governments in the United States. Berkeley:
University of California Press, 1961.

Brosz, Donald J. Establishing Water Conservancy Districts in Wyoming. Bulletin No. 530. Laramie, Wyo.: University of Wyoming, Agricultural
Extension Service, 1970.

Cape, William H.; Graves, Leon B.; and Michaels, Burton M. Government by
Special Districts. Kansas Governmental Research Series No. 37. Lawrence, Kan.: University of Kansas, Governmental Research Center,
1969.

Fischer, Lloyd K. and Timmons, John F. Progress and Problems in the Iowa
Soil Conservation Districts Program: A Pilot Study of the Jasper Soil
Conservation District. Research Bulletin No. 466. Ames, Iowa: Iowa
State College, Agricultural and Home Economics Experiment Station,
1959.

Folmar, Richard. Special District Governments in New Mexico. Santa Fe,
N.M.: Legislative Council Service, 1962.

Geraud, Joseph. Special Taxing Districts in Wyoming. A Report to the Wyoming Legislative Research Committee. Research Report No. 5. Cheyenne, Wyo., 1960.

Institute for Local Self Government. Special Districts or Special Dynasties?
Democracy Diminished. Berkeley: Institute for Local Self Government,
1970.

Makielski, S. J., Jr. and Temple, David G. Special District Government in
Virginia. Charlottesville, Va.: University of Virginia, Institute of Government, 1967.

Missouri. Soil and Water Districts Commission. Organization and Operation
of Soil and Water Conservation Districts in Missouri. Circular No. 1.
Columbia, Mo., 1964.

New Jersey State Chamber of Commerce, Department of Governmental and Economic Research. Government by "Authorities" for New Jersey? Newark, N.J., 1952.

Oregon. State Engineer. Final Report on the Advisability of Creating the Jo-
sephine County People's Utility District. Salem, Ore., 1962.

Park College, Governmental Research Bureau. The Special Districts of Platte
County, Missouri. Parts I and II. Parkville, Mo.: Park College, Governmental Research Bureau, 1958.

Pennsylvania Government Administration Service. Municipal Authorities in
Pennsylvania. Philadelphia: Pennsylvania Government Administration
Service, 1941.

Pock, Max A. Independent Special Districts: A Solution to the Metropolitan Area Problem. Ann Arbor: University of Michigan Law School, 1962.

Preston, Nathaniel Stone. Public Authorities: Devices to Finance, Construct, and Operate Public Projects: How Do They Work? Grass Roots Guides on Democracy and Practical Politics, Booklet No. 40. Washington: Center for Information on America, 1969.

Public Affairs Research Council of Louisiana. Louisiana Levee Districts. Baton Rouge, 1958.

Scott, Stanley and Bollens, John C. Special Districts in California Local Government. Berkeley: University of California, Bureau of Public Administration, 1949.

Scott, Stanley and Corzine, John. Special Districts in the San Francisco Bay Area: Some Problems and Issues. Berkeley: University of California, Institute of Governmental Studies, 1963.

Smith, Robert G. Public Authorities in Urban Areas. Washington: National Association of Counties Research Foundation, 1969.

_____. Public Authorities, Special Districts and Local Government: A Digest of Excerpts. Washington: National Association of Counties Research Foundation, 1964.

Sparlin, Estal E. Special Improvement District Finance in Arkansas. Agricultural Experiment Station Bulletin No. 424. Fayetteville, Ark.: University of Arkansas, College of Agriculture, 1942.

Tax Research Associates of Houston and Harris County, Texas. Harris County Flood Control District: Narrative Functional Description. Houston, Tex., 1952.

Thrombley, Woodward G. Special Districts and Authorities in Texas. Austin, Tex.: University of Texas, Institute of Public Affairs, 1959.

Torgovnik, Efraim. Special Districts in Rhode Island. Metropolitan Study No. 2. Kingston, R.I.: University of Rhode Island, Bureau of Government Research, 1968.

U.S. Department of Agriculture, Economic Research Service. A Selected Bibliography on Special Districts and Authorities in the United States, Annotated. Miscellaneous Publication No. 1087. Washington: U.S. Government Printing Office, 1968.

University of Oregon, Bureau of Municipal Research and Service. Problems of the Urban Fringe. Vols. I and II. Eugene, Ore.: University of Oregon, Bureau of Municipal Research and Service, 1957.

University of Washington, Bureau of Governmental Research and Services. Special Districts in the State of Washington. Report No. 150. Seattle: University of Washington, Bureau of Governmental Research and Services, 1963.

Waters, Harvey Lee and Raines, William H. Special Districts in Kentucky.
    Research Report No. 48. Frankfort, Ky.: Legislative Research Com-
    mission, 1968.

Williams, J. D. A Report on Utah's Special Purpose Districts. Salt Lake City:
    University of Utah, Institute of Government, 1957.

Winter, Arthur B. The Tennessee Utility District: A Problem of Urbanization.
    Knoxville, Tenn.: University of Tennessee Press, 1958.

                            Books--Case Study

Advisory Committee to the Board of Forest Preserve Commissioners. Revised
    Report of Advisory Committee to the Cook County Forest Preserve Com-
    missioners. River Forest, Ill.: Forest Preserve District of Cook
    County, 1959.

Andreas, Alfred Theodore. History of Cook County, Illinois: From the Earli-
    est Period to the Present Time. Chicago: A. T. Andreas, 1884.

Beam, David R., ed. Governing Illinois under the 1970 Constitution. De Kalb,
    Ill.: Northern Illinois University, Center for Governmental Studies,
    1971.

Bishop, Ward L. An Economic Analysis of the Constitutional Restrictions upon
    Municipal Indebtedness in Illinois. Urbana, Ill.: University of Illinois,
    1928.

Board of Forest Preserve Commissioners of Cook County. The Forest Pre-
    serves of Cook County, Illinois. Chicago: Clohesey & Co., 1921.

Braden, George D. and Cohn, Rubin G. The Illinois Constitution: An Anno-
    tated and Comparative Analysis. Urbana, Ill.: University of Illinois,
    Institute of Government and Public Affairs, 1969.

Burnham, Daniel H. and Bennett, Edward H. Plan of Chicago. Chicago: Com-
    mercial Club of Chicago, 1909.

Chamberlin, Everett. Chicago and Its Suburbs. Chicago: T. A. Hungerford &
    Co., 1874.

Chicago Bureau of Public Efficiency. The Nineteen Local Governments in Chi-
    cago: A Multiplicity of Overlapping Taxing Bodies with Many Elective
    Officials. Chicago: Chicago Bureau of Public Efficiency, 1915.

    _____. The Park Governments in Chicago: An Inquiry into Their Organiza-
    tion and Methods of Administration. Chicago: Chicago Bureau of Public
    Efficiency, 1911.

    _____. Unification of Local Governments in Chicago. Chicago: Chicago
    Bureau of Public Efficiency, 1917.

    _____. The Water Works System of the City of Chicago. Chicago: Chicago
    Bureau of Public Efficiency, 1917.

Chicago Home Rule Commission. Modernizing a City Government. Chicago: University of Chicago Press, 1954.

Chicago Park District. First Annual Report: May 1, 1934 to December 31, 1935. Chicago, 1936.

Chicago Urban Transportation District. Annual Report, 1973.

Childs, Mary Louise. Actual Government in Illinois. New York: Century Co., 1914.

Cook County League of Women Voters. The Key to Our Local Government. Chicago: Citizens Information Service, 1966.

Ericson, John. The Water Supply Problem in Relation to the Future Chicago. Chicago, 1925.

Forest Preserve District of Cook County. Land Policy. Rev. ed. River Forest, Ill.: Forest Preserve District of Cook County, 1962.

Fryxell, F. M. The Physiography of the Region of Chicago. Chicago: University of Chicago Press, 1927.

Galloway, James L. and Hetrick, Charles B. Special District Government in Illinois: A Statement of Position. Park Ridge, Ill.: City of Park Ridge, 1962.

Gosnell, Harold F. Machine Politics: Chicago Model. 1937. Reprint. New York: Greenwood Press, 1968.

Greenacre, Alice. A Handbook for the Women Voters of Illinois. Chicago: Chicago School of Civics and Philanthropy, 1913.

Greene, Evarts Boutell. The Government of Illinois: Its History and Administration. New York: Macmillan Company, 1904.

Illinois City Managers' Association. Governmental Structure in the Chicago Metropolitan Area: Facts and Alternatives. Chicago: Cook County Council of the League of Women Voters, 1966.

Johnson, Charles B. Growth of Cook County. Vol. I. Chicago: Board of Commissioners of Cook County, Illinois, 1960.

Kitagawa, Evelyn M. and Taeuber, Karl E., eds. Local Community Fact Book: Chicago Metropolitan Area, 1960. Chicago: University of Chicago, Chicago Community Inventory, 1963.

League of Women Voters of Barrington. In and Around Barrington. N. d.

League of Women Voters of Chicago. Chicago Government: A Reference Book on Local Governmental Units and Agencies in Chicago. Chicago: Citizens Information Service of Metropolitan Chicago, 1954.

League of Women Voters of Chicago Heights. Highlights of Chicago Heights. 1960.

League of Women Voters of Evanston. This Is Evanston. 1970.

League of Women Voters of Glencoe. This Is Glencoe. 1963.

League of Women Voters of Hazel Crest. This Is Hazel Crest. N.d.

League of Women Voters of Hinsdale. This Is Your Community: Hinsdale, Clarendon Hills, Oak Brook. 1967.

League of Women Voters of Homewood and Flossmoor. Know Your Town: Homewood, Flossmoor, Olympia Fields. N.d.

League of Women Voters of Illinois. Structure of Local Government in Illinois: The County. Chicago: League of Women Voters of Illinois, 1968.

_____. Structure of Local Government in Illinois: The Special District. Chicago: League of Women Voters of Illinois, 1969.

_____. Structure of Local Government in Illinois: The Township. Chicago: League of Women Voters of Illinois, 1968.

League of Women Voters of Kenilworth. This Is Kenilworth. 1960.

League of Women Voters of La Grange and La Grange Park. This Is Our Community: La Grange, La Grange Park, Countryside, and La Grange Highlands. 1963.

League of Women Voters of Melrose Park. League's Eye View of Melrose Park. 1962.

League of Women Voters of Morton Grove. This Is Morton Grove. 1970.

League of Women Voters of Northbrook. Northbrook Profile. 1970.

League of Women Voters of Oak Lawn, Illinois. Inside Oak Lawn. N.d.

League of Women Voters of Oak Park and River Forest. Guide to Oak Park and River Forest. 1958.

League of Women Voters of Palos-Orland. Presenting Palos-Orland. 1962.

League of Women Voters of Park Forest. This Is Park Forest. N.d.

League of Women Voters of Riverdale. Spotlight on Riverdale. 1958.

League of Women Voters of Riverdale-Dolton. Dolton: Your Town. N.d.

League of Women Voters of Winnetka-Northfield-Kenilworth. Northfield. N.d.

Lyon, Leverett S., ed. Governmental Problems in the Chicago Metropolitan Area. Chicago: University of Chicago Press, 1957.

Merriam, Charles Edward. Report of an Investigation of the Municipal Revenues of Chicago. Chicago: City Club of Chicago, 1906.

Merriam, Charles E.; Parratt, Spencer D.; and Lepawsky, Albert. The Government of the Metropolitan Region of Chicago. Chicago: University of Chicago Press, 1933.

Metropolitan Sanitary District of Greater Chicago. The Growth and Development of a Modern City. Chicago, 1962.

Northeastern Illinois Metropolitan Area Planning Commission. A Social Geography of Metropolitan Chicago. Chicago: Northeastern Illinois Metropolitan Area Planning Commission, 1960.

Northeastern Illinois Planning Commission. An Introduction to Inter-Governmental Agreements. Chicago: Northeastern Illinois Planning Commission, 1974.

_____. Regional Wastewater Plan: An Element of the Comprehensive General Plan for Northeastern Illinois. Chicago: Northeastern Illinois Planning Commission, 1972.

_____. Sewage Treatment. Metropolitan Planning Guidelines. Phase One: Background Documents. Chicago: Northeastern Illinois Planning Commission, 1965.

_____. Suburban Factbook. Chicago: Northeastern Illinois Planning Commission, 1971.

Platt, Rutherford H. Open Land in Urban Illinois: Roles of the Citizen Advocate. De Kalb, Ill.: Northern Illinois University Press, 1971.

Royko, Mike. Boss: Richard J. Daley of Chicago. Signet Books. New York: New American Library, 1971.

Snider, Clyde F. and Howards, Irving. County Government in Illinois. Carbondale, Ill.: Southern Illinois University, 1960.

Snider, Clyde F. and Anderson, Roy. Local Taxing Units: The Illinois Experience. Urbana, Ill.: University of Illinois, Institute of Government and Public Affairs, 1968.

Solzman, David M. Water Industrial Sites: A Chicago Case Study. Department of Geography Research Paper No. 107. Chicago: University of Chicago, Department of Geography, 1966.

Special Park Commission of Chicago. A Report to the City Council of Chicago on the Subject of a Metropolitan Park System, 1904. Chicago: W. J. Hartman Co., 1905.

Steadman, Robert F. Public Health Organization in the Chicago Region. Chicago: University of Chicago Press, 1930.

Suburban Cook County Tuberculosis Sanitarium District. 20th Anniversary Issue, 1949-1969. Forest Park, Ill.: Suburban Cook County Tuberculosis Sanitarium District, n. d.

United States Public Health Service. The Chicago-Cook County Health Survey. New York: Columbia University Press, 1949.

Walker, Robert Averill. Urban Planning. Chicago: University of Chicago Press, 1941.

Walker, Ward. The Story of the Metropolitan Sanitary District of Greater Chicago. Chicago, 1960.

White, Max R. Water Supply Organization in the Chicago Region. Chicago: University of Chicago Press, 1934.

Periodicals--General

Bollens, John C. "Elements of Successful Annexations." Public Management, XXX (April, 1948), 98-101.

Eyre, John D. "City-County Territorial Competition: The Portsmouth, Virginia Case." Southeastern Geographer, IX (November, 1969), 26-38.

Gilbert, E. W. "The Boundaries of Local Government Areas." Geographical Journal, CXI (April-June, 1948), 172-206.

Hartshorne, Richard. "The Functional Approach to Political Geography." Annals of the Association of American Geographers, XL (March, 1950), 95-130.

Hobday, Victor C. "Should Cities Provide Services to Suburbs or Extend City Limits?" Illinois Municipal Review, XXX (February, 1951), 24.

Jones, Stephen B. "A Unified Field Theory of Political Geography." Annals of the Association of American Geographers, XLIV (June, 1954), 111-23.

Kearns, Kevin G. "On the Nature and Origin of Parks in Urban Areas." Professional Geographer, XX (May, 1968), 167-76.

Melamid, Alexander. "Municipal Quasi-Exclaves: Examples from Yonkers, N.Y." Professional Geographer, XVIII (March, 1966), 94-96.

Murphy, Raymond E. "Town Structure and Urban Concepts in New England." Professional Geographer, XVI (March, 1964), 1-6.

Nelson, Howard J. "The Vernon Area, California--A Study of the Political Factor in Urban Geography." Annals of the Association of American Geographers, XLII (June, 1952), 177-91.

Ostrom, Vincent; Tiebout, Charles M.; and Warren, Robert. "The Organization of Government in Metropolitan Areas: A Theoretical Inquiry." American Political Science Review, LV (December, 1961), 831-42.

Roterus, Victor and Hughes, I. Harding, Jr. "Governmental Problems of Fringe Areas." Public Management, XXX (April, 1948), 94-97.

Taylor, Griffith. "Towns and Townships in Southern Ontario." Economic Geography, XXI (January, 1945), 88-96.

Wehrwein, George S. "The Rural-Urban Fringe." Economic Geography, XVIII (July, 1942), 217-28.

Periodicals--Special Districts

Bird, Frederick L. "The Contribution of Authorities to Efficient Municipal Management." The Authority (December, 1949), pp. 2-4.

Edelstein, Mortimer S. "The Authority Plan--Tool of Modern Government." Cornell Law Quarterly, XXVIII (January, 1943), 177-91.

Gerwig, Robert. "Public Authorities: Legislative Panacea?" Journal of Public Law (January, 1957), pp. 387-88.

"Public Administration Forum: The Special Districts as Instrumentality [sic] of State and Local Government." Midwest Review of Public Administration, I (August, 1967), 119-33.

"The Special District: Taxation Without Representation." Mayor and Manager (July, 1961), pp. 30-32.

"Special Sanitary Districts." The Wisconsin Taxpayer, XXXVIII (October, 1970), entire issue.

Tobin, Austin J. "Administering the Public Authority." Dun's Review (June, 1952), pp. 18-88.

Winslow, David C. "Geographical Implications of Soil Conservation Districts in the United States." Professional Geographer, I (May, 1949), 11-14.

Periodicals--Case Study

Ahlswede, Lee. "Examining Metro Government Today." Illinois County & Township Official, XXXIV (May, 1974), 24-26.

Feldman, A. Daniel. "The Lake Diversion Case--The End of a Cycle." Chicago Bar Record (April, 1968), pp. 270-78.

Ferguson, Harry F. "Sanitary Districts in Illinois." Illinois Municipal Review, III (January-February, 1925), 148-50.

Foreman, Orville N. and Cleary, Edward N. "Municipal Debt Limitation in Illinois." Illinois Municipal Review, XVII (March, 1938), 33-53.

Jones, Victor. "Local Government in Metropolitan Chicago." American Political Science Review, XXX (October, 1936), 935-37.

Kasperson, Roger E. "Toward a Geography of Urban Politics: Chicago, A Case Study." Economic Geography, XLI (April, 1965), 95-107.

Kenny, James B. "Development in Illinois Park Districts: With Special Reference to the Small Park Districts in Cook County." Illinois Municipal Review, XIII (January, 1934), 11-12.

Krausz, N. G. P. "Fire Protection Districts in Illinois." Quarterly of the National Fire Protection Association, XLIV (July, 1950), 178-91.

"Manual for Illinois Park Commissioners/Trustees." Illinois Parks, XII (March-April, 1956), 33.

"Mount Prospect District Approves Master Plan." Illinois Parks, XII (March-April, 1956), 29-32.

"1971 Illinois Association of Park Districts Directory." Illinois Parks and Recreation, II (September-October, 1971), unpaginated.

Olson, W. M. "The Value of Sanitary Districts." The American City, XXVII (December, 1922), 557-63.

Proudfoot, Malcolm J. "Chicago's Fragmented Political Structure." Geographical Review, XLVII (January, 1957), 106-17.

Taylor, Graham Romelyn. "Recreation Developments in Chicago Parks." The Annals of the American Academy of Political and Social Science, XXXV (January-June, 1910), 304-21.

Public Documents--U.S. and Illinois

Illinois. Annotated Statutes.

_____. Constitution (1848).

_____. Constitution (1870).

_____. Constitution (1970).

_____. Laws of Illinois.

_____. Private Laws of the State of Illinois (1869).

Illinois. Commission on Local Government. Report to Governor Richard B. Ogilvie and Members of the 76th Illinois General Assembly. Springfield, Ill., 1969.

Illinois. Geological Survey. Summary of the Geology of the Chicago Area, by H. B. Willman. Circular 460. Urbana, Ill.: Illinois State Geological Survey, 1971.

Illinois. Secretary of State. Counties and Incorporated Municipalities. Springfield, Ill.: issued by Paul Powell, Secretary of State, 1968.

Illinois. Sixth Illinois Constitutional Convention, 1969-1970. Report of Committee on Local Government. Springfield, Ill., n.d.

Illinois. Tax Commission. Survey of Local Finance in Illinois. Vol. I, Atlas of Illinois Taxing Units. Springfield, Ill., 1939.

U.S. Bureau of the Census. Census of Governments: 1962. Vol. V, Local Government in Metropolitan Areas.

_____. Census of Governments: 1967. Vol. 1, Governmental Organization.

U.S. Bureau of the Census. Census of Governments: 1967. Vol. V, Local Government in Metropolitan Areas.

_____. Census of Governments: 1967. Vol. VII, State Reports, No. 13, Illinois.

_____. Census of Governments: 1972. Vol. 1, Governmental Organization.

_____. Census of Population, 1840-1970.

Unpublished Materials and Newspapers

Bixler, William Shelton. "The Government of Cook County, Illinois." Unpublished Master of Philosophy dissertation, Department of History, University of Chicago, 1904.

Board of Forest Preserve Commissioners. "Welcome to Your Forest Preserve District." N.d. (Folder.)

Chicago Daily News. December, 1911; October and November, 1914; and July 26, 1971.

Chicago Tribune. January and February, 1869; May and June, 1889; March and April, 1934.

Christgau, Eugene Frederick. "Unincorporated Communities in Cook County." Unpublished M.A. dissertation, Department of Sociology, University of Chicago, 1942.

City of Chicago, Department of Public Works, Bureau of Maps and Plats. "Map of Chicago: Showing Growth of the City by Annexations and Accretions." 1970.

Cook County. Board of Commissioners. "Map of Cook County, Illinois, 1973, Showing Highways and Forest Preserves."

Cook County. County Clerk. "Comparative Statement of Tax Rates for the Years 1969 and 1970." (Mimeographed.)

Forest Preserve District of Cook County. "Forest Preserve District Comments: Re--Interim Report: Citizens Committee on Cook County Government." Memorandum to Committee on Cook County Government, March 20, 1968. (Mimeographed.)

_____. "The Legislative Enabling Act to Provide for the Creation and Management of Forest Preserve Districts in Illinois." 1960. (Typewritten.)

Hayes, William P. "Development of the Forest Preserve District of Cook County, Illinois." Unpublished M.A. dissertation, Department of History, De Paul University, Chicago, Illinois, 1949.

Illinois. Department of Local Government Affairs. Untitled tax maps of Cook County.

Illinois. Department of Public Works and Buildings, Division of Waterways. "General Drainage Map: Cook County." 1958.

Jensen, Jens Peter. "Financial Statistics of Governments in Cook County." Unpublished report, University of Chicago, 1931.

Klove, Robert Charles. "The Park Ridge-Barrington Area: A Study of Residential Land Patterns and Problems in Suburban Chicago." Unpublished Ph.D. dissertation, University of Chicago, 1942.

Metropolitan Sanitary District of Greater Chicago. "Annexations and Municipalities." 1973. (Map.)

_____. "Map of the Metropolitan Sanitary District of Greater Chicago Showing Sewage Treatment Works, Pumping Stations, Water Reclamation Plants, Retention Reservoirs, Channels, and Sewers." 1971.

Mount Prospect Park District. "Fun Talk." Vol. II (July-August, 1971). Serial issued by the Mount Prospect Park District.

"1972 Park District Task Force Final Report to the Park District of Oak Park." 1973. (Mimeographed.)

Northwest Mosquito Abatement District, Cook County, Illinois. "Report on Mosquito Control Methods, 1971." (Typewritten.)

Oak Leaves (Oak Park, Ill.). January, 1910 to December, 1911.

Olson, Howard E. "Evergreen Park and Mount Greenwood Astride Chicago's Boundary." Unpublished M.A. dissertation, Department of Geography, University of Chicago, 1954.

"Report of the Chicago Mayor's Committee in re City of Chicago-Chicago Park District Consolidation." 1955. (Mimeographed.)

Suburban Cook County Tuberculosis Sanitarium District. "1969 Annual Report." (Folder.)

_____. "Pulmonary Disease Wing." 1969. (Folder.)

THE UNIVERSITY OF CHICAGO
DEPARTMENT OF GEOGRAPHY
RESEARCH PAPERS (Lithographed, 6×9 Inches)

(Available from Department of Geography, The University of Chicago, 5828 S. University Ave., Chicago, Illinois 60637. Price: $5.00 each; by series subscription, $4.00 each.)

84. KANSKY, K. J. *Structure of Transportation Networks: Relationships between Network Geometry and Regional Characteristics* 1963. 155 pp.

91. HILL, A. DAVID. *The Changing Landscape of a Mexican Municipio, Villa Las Rosas, Chiapas* NAS-NRC Foreign Field Research Program Report No. 26. 1964. 121 pp.

94. MC MANIS, DOUGLAS R. *The Initial Evaluation and Utilization of the Illinois Prairies, 1815–1840* 1964. 109 pp.

97. BOWDEN, LEONARD W. *Diffusion of the Decision to Irrigate: Simulation of the Spread of a New Resource Management Practice in the Colorado Northern High Plains* 1965. 146 pp.

98. KATES, ROBERT W. *Industrial Flood Losses: Damage Estimation in the Lehigh Valley* 1965. 76 pp.

102. AHMAD, QAZI. *Indian Cities: Characteristics and Correlates* 1965. 184 pp.

103. BARNUM, H. GARDINER. *Market Centers and Hinterlands in Baden-Württemberg* 1966. 172 pp.

105. SEWELL, W. R. DERRICK, et al. *Human Dimensions of Weather Modification* 1966. 423 pp.

106. SAARINEN, THOMAS F. *Perception of the Drought Hazard on the Great Plains* 1966. 183 pp.

107. SOLZMAN, DAVID M. *Waterway Industrial Sites: A Chicago Case Study* 1967. 138 pp.

108. KASPERSON, ROGER E. *The Dodecanese: Diversity and Unity in Island Politics* 1967. 184 pp.

109. LOWENTHAL, DAVID, et al. *Environmental Perception and Behavior.* 1967. 88 pp.

110. REED, WALLACE E. *Areal Interaction in India: Commodity Flows of the Bengal-Bihar Industrial Area* 1967. 210 pp.

112. BOURNE, LARRY S. *Private Redevelopment of the Central City: Spatial Processes of Structural Change in the City of Toronto* 1967. 199 pp.

113. BRUSH, JOHN E., and GAUTHIER, HOWARD L., JR. *Service Centers and Consumer Trips: Studies on the Philadelphia Metropolitan Fringe* 1968. 182 pp.

114. CLARKSON, JAMES D. *The Cultural Ecology of a Chinese Village: Cameron Highlands, Malaysia* 1968. 174 pp.

115. BURTON, IAN; KATES, ROBERT W.; and SNEAD, RODMAN E. *The Human Ecology of Coastal Flood Hazard in Megalopolis* 1968. 196 pp.

117. WONG, SHUE TUCK. *Perception of Choice and Factors Affecting Industrial Water Supply Decisions in Northeastern Illinois* 1968. 96 pp.

118. JOHNSON, DOUGLAS L. *The Nature of Nomadism* 1969. 200 pp.

119. DIENES, LESLIE. *Locational Factors and Locational Developments in the Soviet Chemical Industry* 1969. 285 pp.

120. MIHELIC, DUSAN. *The Political Element in the Port Geography of Trieste* 1969. 104 pp.

121. BAUMANN, DUANE. *The Recreational Use of Domestic Water Supply Reservoirs: Perception and Choice* 1969. 125 pp.

122. LIND, AULIS O. *Coastal Landforms of Cat Island, Bahamas: A Study of Holocene Accretionary Topography and Sea-Level Change* 1969. 156 pp.

123. WHITNEY, JOSEPH. *China: Area, Administration and Nation Building* 1970. 198 pp.

124. EARICKSON, ROBERT. *The Spatial Behavior of Hospital Patients: A Behavioral Approach to Spatial Interaction in Metropolitan Chicago* 1970. 198 pp.

125. DAY, JOHN C. *Managing the Lower Rio Grande: An Experience in International River Development* 1970. 277 pp.

126. MAC IVER, IAN. *Urban Water Supply Alternatives: Perception and Choice in the Grand Basin, Ontario* 1970. 178 pp.

127. GOHEEN, PETER G. *Victorian Toronto, 1850 to 1900: Pattern and Process of Growth* 1970. 278 pp.

128. GOOD, CHARLES M. *Rural Markets and Trade in East Africa* 1970. 252 pp.

129. MEYER, DAVID R. *Spatial Variation of Black Urban Households* 1970. 127 pp.

130. GLADFELTER, BRUCE. *Meseta and Campiña Landforms in Central Spain: A Geomorphology of the Alto Henares Basin* 1971. 204 pp.

131. NEILS, ELAINE M. *Reservation to City: Indian Urbanization and Federal Relocation* 1971. 200 pp.

132. MOLINE, NORMAN T. *Mobility and the Small Town, 1900–1930* 1971. 169 pp.

133. SCHWIND, PAUL J. *Migration and Regional Development in the United States, 1950–1960* 1971. 170 pp.

134. PYLE, GERALD F. *Heart Disease, Cancer and Stroke in Chicago: A Geographical Analysis with Facilities Plans for 1980* 1971. 292 pp.

135. JOHNSON, JAMES F. *Renovated Waste Water: An Alternative Source of Municipal Water Supply in the U.S.* 1971. 155 pp.

136. BUTZER, KARL W. *Recent History of an Ethiopian Delta: The Omo River and the Level of Lake Rudolf* 1971. 184 pp.

137. HARRIS, CHAUNCY D. *Annotated World List of Selected Current Geographical Serials in English, French, and German* 3rd edition 1971. 77 pp.

138. HARRIS, CHAUNCY D., and FELLMANN, JEROME D. *International List of Geographical Serials* 2nd edition 1971. 267 pp.

139. MC MANIS, DOUGLAS R. *European Impressions of the New England Coast, 1497–1620* 1972. 147 pp.

140. COHEN, YEHOSHUA S. *Diffusion of an Innovation in an Urban System: The Spread of Planned Regional Shopping Centers in the United States, 1949–1968* 1972. 136 pp.

141. MITCHELL, NORA. *The Indian Hill-Station: Kodaikanal* 1972. 199 pp.

142. PLATT, RUTHERFORD H. *The Open Space Decision Process: Spatial Allocation of Costs and Benefits* 1972. 189 pp.

143. GOLANT, STEPHEN M. *The Residential Location and Spatial Behavior of the Elderly: A Canadian Example* 1972. 226 pp.

144. PANNELL, CLIFTON W. *T'ai-chung, T'ai-wan: Structure and Function* 1973. 200 pp.

145. LANKFORD, PHILIP M. *Regional Incomes in the United States, 1929–1967: Level, Distribution, Stability, and Growth* 1972. 137 pp.

146. FREEMAN, DONALD B. *International Trade, Migration, and Capital Flows: A Quantitative Analysis of Spatial Economic Interaction* 1973. 202 pp.

147. MYERS, SARAH K. *Language Shift Among Migrants to Lima, Peru* 1973. 204 pp.

148. JOHNSON, DOUGLAS L. *Jabal al-Akhdar, Cyrenaica: An Historical Geography of Settlement and Livelihood* 1973. 240 pp.

149. YEUNG, YUE-MAN. *National Development Policy and Urban Transformation in Singapore: A Study of Public Housing and the Marketing System* 1973. 204 pp.

150. HALL, FRED L. *Location Criteria for High Schools: Student Transportation and Racial Integration* 1973. 156 pp.

151. ROSENBERG, TERRY J. *Residence, Employment, and Mobility of Puerto Ricans in New York City* 1974. 230 pp.

152. MIKESELL, MARVIN W., editor. *Geographers Abroad: Essays on the Problems and Prospects of Research in Foreign Areas* 1973. 296 pp.

153. OSBORN, JAMES. *Area, Development Policy, and the Middle City in Malaysia* 1974. 273 pp.

154. WACHT, WALTER F. *The Domestic Air Transportation Network of the United States* 1974. 98 pp.

155. BERRY, BRIAN J. L., et al. *Land Use, Urban Form and Environmental Quality* 1974. 464 pp.

156. MITCHELL, JAMES K. *Community Response to Coastal Erosion: Individual and Collective Adjustments to Hazard on the Atlantic Shore* 1974. 209 pp.

157. COOK, GILLIAN P. *Spatial Dynamics of Business Growth in the Witwatersrand* 1975. 143 pp.

158. STARR, JOHN T., JR. *The Evolution of Unit Train Operations in the United States: 1960–1969—A Decade of Experience* 1975.

159. PYLE, GERALD F. *The Spatial Dynamics of Crime* 1974. 220 pp.

160. MEYER, JUDITH W. *Diffusion of an American Montessori Education* 1975. 109 pp.

161. SCHMID, JAMES A. *Urban Vegetation: A Review and Chicago Case Study* 1975.

162. LAMB, RICHARD. *Metropolitan Impacts on Rural America* 1975.

163. FEDOR, THOMAS. *Patterns of Urban Growth in the Russian Empire during the Nineteenth Century* 1975.

164. HARRIS, CHAUNCY D. *Guide to Geographical Bibliographies and Reference Works in Russian or on the Soviet Union* 1975. 496 pp.

165. JONES, DONALD W. *Migration and Urban Unemployment in Dualistic Economic Development* 1975.

166. BEDNARZ, ROBERT S. *The Effect of Air Pollution on Property Value* 1975. 118 pp.

167. HANNEMANN, MANFRED. *The Diffusion of the Reformation in Southwestern Germany, 1518-1534* 1975.

168. SUBLETT, MICHAEL D. *Farmers on the Road. Interfarm Migration and the Farming of Noncontiguous Lands in Three Midwestern Townships, 1939-1969* 1975. 228 pp.

169. STETZER, DONALD FOSTER. *Special Districts in Cook County: Toward a Geography of Local Government* 1975. 189 pp.